Coworking Spaces

Janet Merkel • Dimitris Pettas • Vasilis Avdikos
Editors

Coworking Spaces

Alternative Topologies and Transformative Potentials

Springer

Editors
Janet Merkel
Institute of Urban and Regional Planning
Technische Universität Berlin
Berlin, Germany

Dimitris Pettas
Institute of Urban and Regional Planning
Technische Universität Berlin
Berlin, Germany

Vasilis Avdikos
Department of Economic and Regional Development
Panteion University
Athens, Greece

ISBN 978-3-031-42267-6 ISBN 978-3-031-42268-3 (eBook)
https://doi.org/10.1007/978-3-031-42268-3

This Springer imprint is published by the registered company Springer Nature Switzerland AG
The registered company address is: Gewerbestrasse 11, 6330 Cham, Switzerland

Paper in this product is recyclable.

Acknowledgements

We wish to thank all authors who worked with us in this book and who were willing to respond to our feedback and critique during the editing process, but as well as through an online seminar in the summer of 2022 and during the European Conference "The future of collaborative workspaces: a multi-scalar and interdisciplinary approach", held in Athens in January 2023. We also thank all scholars who joined us on several sessions at International Critical Geography Conference (ICCG) in Athens 2019, Research Committee 21 on Sociology of Urban and Regional Development of (RC21) in Antwerp in 2021 and Athens in 2022, and at the American Associations of Geography (AAG) annual conference in Seattle in 2021. Those discussions held online and in person helped us to develop most of the ideas on coworking that take shape in this collection. Moreover, we would like to warmly thank Ashley Proctor and Bernie Mitchell for giving us an interview that appears in the last section of the book. We are grateful for the support of our work colleagues at Technical University Berlin and Panteion University of Social and Political Sciences in Athens and the funding of the Hellenic Foundation for Research and Innovation for the Urban Cowork project that made some of these sessions possible. We are also grateful to our colleagues in the COST Action network "The geography of New Working Spaces and the impact on the periphery" (CA18214) that involves more than 140 research partners from 35 countries. For the past four years, the network provided a crucial meeting space for researchers in this young and interdisciplinary research field. With the creation of the new Marie Sklodowska Curie Innovative Training Network CORAL-ITN, these conversations and shared research interests find a continuation. At Springer, our thanks go to Zachary Romano, Senior Editor, for supporting the original idea for the book and, with his colleague, Herbert Moses, steering us through the editorial process. We are also grateful to the proposal reviewers who provided important encouragement with their helpful comments. The UrbanCoWork research project was supported by the Hellenic Foundation for Research and Innovation (H.F.R.I.) under the "1st Call for H.F.R.I. Research Projects to support Faculty Members & Researchers and the Procurement of High-and the procurement of high-cost research equipment grant" (Project Number: 1932).

Contents

Coworking Spaces: Alternative Topologies and Transformative Potentials 1
Janet Merkel, Vasilis Avdikos, and Dimitris Pettas

Part I Coworking and the Politics of Self-Organization and Community Organization in the City

***A Place in the World*: A Mapping of Coworking as a Site/Practice of Commoning and Community in İstanbul** 17
Özlem İlyas

Self-Organised Labour: Reclaiming the Commons in the Post-Socialist Space of Southeast Europe 33
Dimitra Gkitsa

Organisational Decline and the Failure in Alternative Organising: The Case of a Coworking Cooperative 51
Eeva Houtbeckers and Maja Korica

CWS as Terrains for the (De)territorialisation of Labour: Digital Nomads, Local Coworkers and the Pursuit for Resilient and Cooperative Spaces 67
Dimitris Pettas and Vasilis Avdikos

Part II Care and Labour in Coworking Practices

Coworking Spaces as Social Infrastructures of Care 83
Janet Merkel

Coworking Spaces in Aotearoa New Zealand: Embodied Geographies of Care in Pandemic Times 97
Joke J. Methorst

You Know That You've Succeeded in Your Role When Your Work Renders You Invisible: The Invisible Work of Community Management . 111
Antigoni Papageorgiou and Martha Michailidou

Exhausting Coworking: On the Implications of Reproductive Work for Coworkers' Subjectivities 125
Nikos Gatsinos and Malte Höfner

Part III Alternative Modes of Development Through Coworking and Policy Responses

The Abundant Crops of Subjectivisation and the Difficult Arts of Institutionalisation: Building Transformative Collaborative Spaces of Urban Labour in the Context of the Mares de Madrid Initiative 141
Alessandro Coppola

Public Support for New Working Spaces in Italy: The Case of Presidi di Comunità 159
Ilaria Mariotti, Ivana Pais, and Francesca Ciccarelli

Exploring Social Innovative Collaborative Workspace Assemblages: A Comparative Study from Athens 175
Colm Stockdale and Lorenzo Marmo

Facilitating Social Economy in Taiwan: A Case Study of a University-Community Coworking Space in an Indigenous Community of Eastern Taiwan 191
Muyi Chou and Salizan Takisvilainan

Spaces of Collaboration-Spaces of Possibility: How a Local Hub Creates New Possibilities in a Greek Peripheral Town 207
Danai Liodaki

Part IV Epilogue

A Conversation with Ashley Proctor and Bernie Mitchell 221
Janet Merkel, Dimitris Pettas, and Vasilis Avdikos

Index 241

Contributors

Vasilis Avdikos is an associate professor at the Department of Economic and Regional Development, Panteion University of Social and Political Sciences since 2014. He holds a doctorate from the University of Sheffield and an MSc degree from the University of Strathclyde. His research interests include urban and regional development, and the cultural and creative industries. He has led and coordinated several national and international research projects; currently he is researching the impacts of collaborative workspaces in rural areas (MSCA CORAL-ITN) and the ways commons can offer new and sustainable solutions in GLAMs (GLAMMONS-Horizon Europe). He is the author of two monographs and two other collective volumes and several research papers.

Muyi Chou is an assistant professor at the bachelor's degree program in Ocean Tourism Management, National Taiwan Ocean University (NTOU). She obtained her PhD in Political Science from Humboldt University of Berlin. Then she served as a postdoctoral research fellow at the Department of Sociology, National Taiwan University and participated in Humanity Innovation and Social Practice Project. Her research interests include civil society, social and solidarity economy, deliberative democracy, university-community partnership research and social innovation.

Francesca Ciccarelli is a PhD candidate at the Department of Architecture and Urban Studies of the Politecnico di Milano and early-stage researcher at the CORAL-ITN network (2021-2024). Her current research explores the connection between workspaces, peripheral regions and workers' wellbeing and quality of life.

Alessandro Coppola is an associate professor of Urban Planning at the Department of Architecture and Urban Studies, Politecnico di Milano. His research interests include neighbourhood policy and politics, urban governance, housing policy, housing informality and urban shrinkage. He currently coordinates research projects on residential mobility, housing financialisation and area-based policies for internal areas. His contributions have appeared in journals including *Cities*, *Geografiska Annalers*, *Progress in Planning*, *Urban Studies*, *Urban Geography* and

International Journal of Urban Sustainable Development. He has been a visiting fellow at Johns Hopkins University and Cleveland State University, University of Coimbra, Mistra Urban Futures in Goteborg, and the Leibniz Institutes for Geography in Erkner and Leipzig, Sciences Po.

Nikos Gatsinos is a Marie Skłodowska-Curie fellow and PhD candidate at the Institute of Geography and Regional Science, University of Graz, Austria. After having completed his bachelor's degree in sociology at Panteion University of Athens, he pursued a master's degree in Social and Solidarity Economy at the Hellenic Open University. His past working experience was in the field of asylum, where he worked with asylum seekers as a case officer for the Greek Asylum Service applying qualitative methods. His research interests focus on workers' agency and the affective dimensions of social interaction which he has previously explored in the context of the Athenian co-working landscape. Following on from his previous research, in his role within CORAL-ITN, he addresses the issue of precariousness and the ways it is negotiated through collaborative practices in rural and peripheral areas.

Dimitra Gkitsa is a cultural practitioner and researcher working in the intersection of memory studies, affect, contemporary art practices and the post-socialist visual cultures. Holding a PhD in Visual Cultures from Goldsmiths University of London, she is currently the Alexander Nash Research Fellow at the School of Slavonic and East European Studies, University College London (UCL). Previously, she was an associate lecturer in Visual Cultures at Goldsmiths, University of London, and a module leader in Fine Arts (Critical and Cultural Studies) at the University of Hertfordshire. Outside academia, she has worked with numerous cultural and arts organisations, including the British Council in Athens and the Institute of Contemporary Arts (ICA) in London.

Malte Höfner is a research assistant and a PhD candidate at the Institute of Geography and Regional Science, University of Graz, Austria. After completing his undergraduate degree in geography at the University of Marburg, Germany, he worked on the topic of coworking as a collective coping strategy for boundaryless work of creative labour in his master's degree in sustainable urban and regional development at the University of Graz. Meanwhile, he was actively involved in mediating vacant spaces for temporary use at a project called 'Raumbasis'. In his dissertation project 'Raumteilen' (*Sharing Spaces*), he deals with different geographies of sharing looking into everyday domains such as spaces of work and investigates on how social practices of sharing influence, affect and (re)produce the spheres of everyday life for the subjects participating in sharing. Approaching the geographies of collaborative workspaces through the lens of practice theory, he pays attention to different forms of work that become visible through sharing. With a focus on qualitative social research, he teaches methods in human geography at the University of Graz.

Eeva Houtbeckers (DSc Econ. & Buss.Admin.) studies economies, work and livelihoods during sustainability transformations. Her postdoctoral ethnographic study about postgrowth work practices in the global north contributes to the growing movement that challenges economic growth as an imperative. Inspired by ecofeminist thinking, she participated in and followed projects on degrowth, self-sufficiency and forest and land management. She works as a senior researcher at the University of Turku (2023–2024) in a project about post-fossil working life. She is co-founder of Untame research collective and *Degrowth* journal.

Özlem İlyas is a PhD candidate in Modern Turkish History at Boğaziçi University, Istanbul. She completed her MA in Sociology at the same university with a dissertation on the working and living conditions of freelancers from different industries in Turkey. This work was translated into and published as a book in Turkish under the title *Freelance Emek: Ofissiz Çalışmanın Sınıfsallığı* [Freelance Labour: Classed Nature of Officeless Work] by İletişim Yayınları in 2021. Her PhD research will scrutinise the relation between the transformation of higher education and labour markets in Turkey. She is a member of the Association of Economic and Social Analysis and is currently a board member of *Rethinking Marxism: A Journal of Economics, Culture & Society*. Her research interests revolve around precarious work, Marxist theory, psychoanalysis, youth studies and labour ethnographies.

Maja Korica holds the senior academic title of Reader in Management and Organisation at Warwick Business School, University of Warwick. She specialises in studying the work of senior leaders and boards of directors, and the consequences of their actions on workplaces, neighbourhoods and societies. She is the founding organiser of the Interdisciplinary Studies of Elites reading group, which launched in October 2022.

Danai Liodaki studied architecture and research in architecture, space and culture in the National Technical University of Athens in Greece. She has worked in research projects on issues such as participation, the commons, social movements for the environment and the right to the city. She has also worked as a development consultant in collaboration with municipalities and regional governments in Greece and in the cultural sector in Greece and Germany. She is currently a PhD candidate in Human Geography at the University of Leipzig, working in the Leibniz Institute for Regional Geography, as a Marie Skłodowska-Curie Doctoral Fellow in the framework of CORAL Innovative Training Network.

Ilaria Mariotti is an associate professor of Urban and Regional Economics at the Department of Architecture and Urban Studies – Politecnico di Milano (Italy). She received an MSc degree in Regional Studies at the University of Reading (UK), a PhD in Economic Geography at the University of Groningen (NL) and a PhD in Transport Economics at the Università degli Studi di Genova (IT). She teaches urban and regional economics and environmental economics in undergraduate, graduate and PhD courses. She is member of the Urban Planning, Design and Policy

PhD programme at Politecnico di Milano. Her main research interests concern: firm location, industrial districts and clusters; transport and logistics industry; new working spaces and their effects on the local context. She is Chair of the Cost Action 18214 "The Geography of New Working Spaces and the Impact on the Periphery" (2019-2024), and Project coordinator for the Politecnico di Milano team of the CORAL "Exploring the impacts of collaborative workspaces in rural and peripheral areas in the EU" Project, Horizon 2020 – Marie Curie – ITN-2020 (Marie Skłodowska-Curie Innovative Training Networks) (2021-2024).

Lorenzo Marmo is an early-stage researcher in the Marie Sklodowska-Curie Innovative Training Networks (CORAL-ITN), hosted by the European Creative Hubs Network. He is a PhD candidate at the Faculty of Regional Development, Panteion University, Athens. He has a background in development economics and sustainable agricultural practices. During his master's degree in management, economics and consumer studies at Wageningen University (the Netherlands), he acquired knowledge in agricultural and rural development, climate change and smallholders' food security in Least Developed Countries. During his experiences as research intern at CGIAR and as project intern in the Development Economics Group of WUR, he had the opportunity to strengthen his econometrics and quantitative research skills. Because of the importance he places in new opportunities for socio-economic development in disadvantaged and rural areas, his professional ambition is to shed light on latent dynamics and impacts of Collaborative Workspaces, so as to efficiently respond to the specific local and regional needs and pave the way for a more sustainable future.

Janet Merkel is an urban sociologist based at the Institute of Urban and Regional Planning at Technical University of Berlin where she works on the social and spatial re-organisation of independent modes of work. She holds a PhD in Urban and Regional Sociology from Humboldt University Berlin. Her research interests include creative industries and creative labour, cultural planning and urban cultural policy. She has held positions at Berlin's Social Science Research Center (WZB) and Hertie School of Governance and was working as a lecturer for Culture and Creative Industries at City, University of London and guest professor of Economics of Urban and Regional Development at the University of Kassel.

Joke J. Methorst is a PhD researcher at the Department of Geography and Te Ngira: Institute for Population Research at the University of Waikato in New Zealand, where she is also a recipient of the University of Waikato Doctoral Scholarship. Her PhD research is centred around the changing digital geographies of work, exploring the socio-spatial dynamics of location-independent work through a focus on bodies, mobility, technology and place. Joke's background lies in tourism and environment, as she holds an MSc in tourism, society and environment from the Cultural Geography Group of Wageningen University in the Netherlands. Her research interests centre around transnational mobility, the production of place/

space, geographies of care and the socio-spatial impact of technologies on the geographies of work, mobility and place.

Martha Michailidou is an assistant professor in Methods of Communication Research at the Department of Communications, Media and Culture, Panteion University in Athens. She is currently a member of two research projects researching the impacts of collaborative workspaces in rural areas (MSCA CORAL-ITN) and the ways commons can offer new and sustainable solutions in GLAMs (GLAMMONS-Horizon Europe). Her research interests include labour in the creative/cultural industries, gender in the creative/cultural industries, multi-methodological approaches to media and cultural research, and feminist, emergent and digital methods.

Ivana Pais is a professor of Economic Sociology at Università Cattolica del Sacro Cuore. Her research interests focus on platform economy and digital labour. She is the director of TRAILab – Transformative Actions Interdisciplinary Laboratory at Università Cattolica del Sacro Cuore. She is co-editor-in-chief of *Sociologica.* International Journal for Sociological Debate. Her main research interests concern: digital labour, platform forms of organisation and sociology of economic innovation. She is the principal investigator of the WePlat project (Welfare systems in the age of platforms: drivers of change for users, providers and policy makers).

Antigoni Papageorgiou is a post-doctoral researcher at the Department of Economic and Regional Development, Panteion University of Social and Political Sciences. Her main research interests include but are not limited to cultural work, gender and the production of subjectivity in the cultural and creative industries. She has a background in Political Analysis (BA in Political Science and Public Administration, National Kapodistrian University of Athens) and Cultural Management (MSc in Cultural Management, Panteion University of Political and Social Sciences). In 2020, she received her PhD in Cultural Industries funded by the Arts and Humanities Research Council (AHRC) from the University of Leeds when she successfully defended her thesis titled 'Managing the Start-up Working Life: The Emergence of Creative Labour, Coworking Spaces and Hubs in Athens'. She is specialised in qualitative research and ethnographic techniques. In recent years, she has published on the emerging forms of subjectivities, the significance of creative labour, the gendered aspects of entrepreneurial labour and the sexist practices that penetrate traditional cultural industries as well as the wider creative sectors.

Dimitris Pettas is a Marie Skłodowska-Curie Postdoctoral Fellow at the Institute of Urban and Regional Planning, Technical University of Berlin (TUB). He holds a PhD from the National Technical University of Athens (School of Architecture). He has worked as a postdoctoral researcher at the Regional Development Institute (Panteion University), the National Technical University of Athens and the Research Centre for the Humanities. He has also been a lecturer in the Social and Solidarity Economy postgraduate programme at the Hellenic Open University. His research interests include the study of: the social production of public space, the development of platform economy and its impact on urban environment, the role of power relations and agency in modes of urban governance, the emergence and broader transformative potentialities of grassroots collaborative and social/solidarity economy initiatives, and the exploration and comparative analysis of epistemological and ontological approaches concerning the production of urban environment and urban life.

Colm Stockdale is a PhD candidate at the Department of Economic and Regional Development, Panteion University, Athens, and is part of the Marie Skłodowska-Curie Innovative Training Network (CORAL-ITN) where he is researching how collaborative workspaces are involved in social innovation process in rural and peripheral areas. He has a master's degree in Co-operatives, Agri-Food and Sustainable Development from University College Cork. His research interests lie in social innovation, grassroots initiatives in the social and solidarity economy, and the transformative potential of these towards broader socio-ecological transitions.

Salizan Takisvilainan is from *Nakahila* in Hualien County, Taiwan. He is a poet and a writer. He writes about his love and sorrow for his hometown and his concern with the land and ethnic culture. He is the founder of Bunch of Millet Bunun Independent Publishing Studio and devotes himself to preserving Bunun language and culture. He has won many important literature prizes in Taiwan and published four books. One of his books *Carrying Mountains by the Headband: Stories of Mountain Guides and Rangers*(用頭帶背起一座座山:嚮導背工與巡山員的故事 is translated in Czech version.

Coworking Spaces: Alternative Topologies and Transformative Potentials

Janet Merkel, Vasilis Avdikos, and Dimitris Pettas

Since Clay Spinuzzi's "Working Alone Together" (2012), which is considered the first academic publication on coworking, multidisciplinary scholarship on coworking as a socio-spatial practice has developed a rich body of empirical research over the last decade. Researchers from many academic disciplines have provided the coworking literature with multiple accounts, exploring the diverse functions and impacts of coworking spaces. Yet a growing concern in academic debates is the lack of theorization and coworking research is often exposed to a critique of being overly descriptive, and so it merely fails to fully grasp and unpack the processes that make coworking a transformative medium of social and economic relations. Although the focus and the nucleus of coworking spaces is and should continue to be labor, the dynamics of coworking have multiple manifestations that intersect with various processes of spatial development.

With this collection we aim to bring together authors that provide different theoretical or conceptual lenses on coworking. From coworking as a commoning practice and as an alternative medium of community organizing, to coworking as care and as mediator of social innovation practices, the 13 chapters that follow attempt to ground (theoretically and empirically) the discussions about the transformative power of coworking and bring to the fore the new potentials that coworking can offer to workers, communities, and cities.

When we embarked on the journey of creating an edited book on coworking our aims have been twofold: First, we want to make visible the plurality of existing practices around shared resources in coworking. We were especially interested in making visible the assemblages of human and non-human actors as agents of change

J. Merkel (✉) · D. Pettas
Institute of Urban and Regional Planning, Technische Universität Berlin, Berlin, Germany
e-mail: janet.merkel@tu-berlin.de

V. Avdikos
Department of Economic and Regional Development, Panteion University, Athens, Greece

J. Merkel et al. (eds.), *Coworking Spaces*, https://doi.org/10.1007/978-3-031-42268-3_1

associated with coworking and the re-organization of work and labor power. Second, we aim to develop an emancipatory narrative for coworking and discuss in-depth the role of coworking spaces for coworkers but also for neighborhoods, cities and beyond. A narrative that does not emphasize entrepreneurship or coworking as the epitome of the "neoliberal entrepreneurial self" as in the dominant interpretations in the current research, but rather one that centers coworking around the creation of meaningful, care-full social relationships, supporting empathy and an ethics that recognizes mutual interdependencies and builds a strong foundation for social change. So, it is about alternative narratives, emancipation politics and the wider social role that coworking and coworking spaces might play in different socio-spatial contexts such as neighborhoods, cities, regions, or rural areas because they constitute crucial contexts for the formation and maintenance of social relationships and transformative potentials. We acknowledge that a variety of spaces could be explored as sites at which coworking can be practiced, and where potentials of coworking might be explored. Those can be maker spaces, hackerspaces, libraries or even more transient and less dedicated spaces such as cafés, hotel lobbies, stations, or private homes (Merkel, 2022; Reuschke et al., 2021; Shearmur et al., 2021). In our own research, however, we focus on small, independent, and often community-led coworking spaces that are mainly frequented by freelance workers and invited contributions for this volume that similarly investigate the potentials of smaller spaces.

For us, coworking practices extend far beyond the bounded limits of a coworking space (as physical space). Instead, coworking spaces operate as the terrains for the territorialization and channeling of multi-leveled flows of people, capital, knowledge, innovation, and services. Building on an anti-essentialist, relational understanding of urban processes, we conceptualize coworking topologies as complex networks, formulated by components which are not defined by a specific essence (translated in predefined behaviors, capacities, and roles) but, instead, by the continuous negotiations, alterations and adaptations that are driven by entering specific sets of relations. Given the wide range of coworking attributes, motivations, sizes, organizational modes, etc., such a topological thinking and understanding provides us with the theoretical and methodological flexibility to constantly re-evaluate the role and broader outreach of coworking practices, and especially the interconnectedness and the interdependencies of coworking nowadays. Moreover, under the condition of increased global connectivity concerning communication, transportation, economic and cultural flows, coworking spaces' networks impact and outreach transcend pre-defined scales of reference. From engaging with local communities and addressing the needs of local freelancers and professionals, to their participation in global coworking networks and attracting remote workers from all over the globe, coworking practices and networks are unfolding in different geographical scales.

We hope that this collection will, first, provide theoretical and empirical analysis to scholars working in the field of coworking research. With information-rich case studies and new ways of theorizing and conceptualizing coworking, we want to contribute to the development of this young multidisciplinary research field. A second way is to read the book as contributing towards broader debates on new ways of working, new geographies and spatialities of work and in "workers' abilities to produce and manipulate geographic space" (Herod, 1997, p. 3). With the ongoing digitalization and the shift towards remote work since the COVID-19 pandemic, it has become more urgent to provide analysis that brings out the multi-locational character of work across different sites (e.g., from home, coworking spaces, cafés, cars, libraries, etc.) and traces the everyday geographies of work and the emerging new "workscapes" of independent and increasingly remote workers (Brakel-Ahmed et al., 2020; Stevens & Shearmur, 2020). Moreover, the policy responses during the COVID-19 pandemic have exposed how this expanding field of atypical and independent modes of work is still not well understood and left them out of emergency funds and support packages (de Peuter et al., 2022; Pulignano et al., 2021). Here, coworking spaces can be a valuable prism to study the "absences and ambiguities in the freelancing labour relation" (Worth & Karaagac, 2021). A third use speaks to those who engage in debates of emancipatory and transformative potentials. We hope our case studies contribute useful insights and perspectives for practitioners and academics to debates on local organizing, alternative social and economic arrangements, etc.

The Emergence of Coworking and Coworking Spaces

Coworking as a socio-spatial practice of working in a shared office space with other unaffiliated workers has evolved with the rise of new independent and nonstandard modes of work such as freelance or self-employment over the last two decades (Boeri et al., 2020; Kalleberg, 2018; Murgia & Pulignano, 2021; Worth & Karaagac, 2021). The increase of independent modes of work is a result of the ongoing deregulation and flexibilization of labor (Kalleberg, 2018) and the spatial reorganization of work (Felstead et al., 2005) due to ever-expanding information and communication technologies with constantly improving computing technologies, wireless communication devices, and data storage systems that facilitate location independent, mobile work (Stevens & Shearmur, 2020) and the rapidly expanding gig-based platform work (Vallas & Schor, 2020). The rise of remote work, accelerated through the home office experience for many employees during the COVID-19 pandemic, additionally now fuels the development of coworking spaces (Clifton & Reuschke, 2022; Mariotti et al., 2022).

While shared workspaces for artists, small business and creative practitioners have a much longer history (Montgomery, 2007), the financial crisis 2007–2008 and the following economic recession that saw millions of people lose their jobs, accelerated independent modes of work and the growth of coworking spaces. Over the last decade coworking has become diversified, commercialized, financialized, and integrated into neoliberal urban and state's entrepreneurial polices (Luo & Chan, 2020; McNeill, 2021; Moisio & Rossi, 2020; Schmidt et al., 2014). Moreover, coworking also spread into different spatial contexts beyond the metropolitan cores of big cities (Mariotti et al., 2021). Increasingly, smaller and more community-led coworking spaces now grow outside big cities in rural and peripheral areas (see, e.g., for Germany Bähr et al., 2021; Capdevila, 2021).

The multidisciplinary research on coworking has mainly focused on understanding why coworkers choose to cowork, the kind of resources these spaces provide coworkers with such as network formation and supporting entrepreneurial development and innovation (Bouncken et al., 2021; Boutillier et al., 2020; Butcher, 2018; Grazian, 2020; Howell, 2022; Kraus et al., 2022; Rådman et al., 2022; Spinuzzi et al., 2019). Much less attention has been paid how coworking spaces, despite being framed as "open" and "inclusive", might reproduce existing social inequalities around gender, class, or race (Fast, 2022; Lorne, 2020; Richardson, 2015). And, despite a sustained growth of coworking spaces in Asia and Africa, there is little research examining coworking spaces outside the North American and European contexts (England et al., 2021; Luo & Chan, 2020; Tintiangko & Soriano, 2020).

Coworking signifies a social and spatial reorganization of work that now affects a growing number of professional activities, occupations, and sectors. While independent workers choose to work from a shared workspace, many employees nowadays must also work in shared spaces as employers remodel existing office space after coworking spaces or because companies move into shared workspaces to save on facility management (Harris, 2021). Moreover, the COVID-19 pandemic has accelerated employees' preferences for remote work options. The proximity of the workplace to the home has now become an important criterion for many remote workers (Clifton & Reuschke, 2022; Shearmur et al., 2021). Yet, even though many employees want to work closer to their home, they do not necessarily want to work from home and prefer shared workspaces. Subsequently, coworking spaces are now expanding into suburban areas, shopping centers or small and medium-sized cities (Krasilnikova & Levin-Keitel, 2022). Thus, coworking and coworking spaces as socio-spatial phenomena signify major shifts in the spatial re-organization of work, work mobility and the geography or workplaces.

Framing Key Themes: Overview of the Book

We use three different themes as lenses to think through coworking topologies and the emancipatory and transformative potentials of coworking. The first one we see at the core of the coworking praxis: that is labor and the use of resources in coworking arrangements. The second consists of the evolution of a new kind of ethics in coworking spaces that is the ethics of care and solidarity amongst coworkers and hosts. The third one aims to unpack the broader contribution of coworking in boosting alternative modes of development in the city, and the ways that specific public policies have assisted (or not) and in what ways coworking spaces in doing so.

Coworking and the Politics of Self-Organization and Community Organization in the City

Our understanding of the emergence of coworking spaces is grounded in the ongoing transformations of work and employment relationships under capitalist development. Coworking spaces provide independent workers such as freelance and self-employed workers with crucial resources for conducting their work such as knowledge (Smith, 2020), network contacts (Gandini, 2015), sociality (Garrett et al., 2017), professional identity (Bacevice & Spreitzer, 2022), reputation (Gandini et al., 2017) or affect and atmosphere (Waters-Lynch & Duff, 2019). The lack of an organizational work integration for these independent modes of work has led freelance workers to create new social and spatial relations—the first space as it is told in coworking research was found by freelance journalists in 2005 in San Francisco (Howell, 2022, p. 4). Coworking spaces have altered the ways resources are organized and managed between independent workers. Subsequently, labor relations have also been rearranged along with the ways labor precariousness is manifested and counterbalanced through community practices and strategies. Research on coworking has shown for example that coworking can serve as identity anchoring environments for flexible workers to help "shape their professional identities while providing other important attributes of work to help it feel embodied and grounded" (Bacevice & Spreitzer, 2022). For example, labor geography has shown that the ability to shape the conditions of their working lives to both survive and thrive differs greatly for people in different places, and in different sectors of society. And as Andrew Herold remarked, "workers' abilities to produce and manipulate geographic space in particular ways is a potent form of social power" (Herod, 1997, p. 3).

In this perspective, the contributions in this section look at the intersection of self-organization practices and community approaches over the management and use of shared resources in coworking. For a better understanding of the operation and the defining forces that materialize the coworking praxis and the broader transformative impact of coworking spaces, we augment previous discussions (see, e.g., Gandini & Cossu, 2021; Smith, 2020; Vidaillet & Bousalham, 2020) on the potentialities of coworking spaces through a wide range of contributions.

The commons framework, as well as research on the cooperative movement and the landscape of Social and Solidarity Economy, may provide us with the departure points for elaborating the construction, use and management over and of resources in coworking spaces and the social arrangements around it. Building on recent scholarship on the commons and commoning, we understand the commons as a way of collective self-organization outside capitalist logic (Huron, 2015), but still inside capitalist governmentalities. Commoning, as a social process and collective activity, is a way of socializing resources needed to sustain life (Linebaugh, 2008) and to foster self-organization within communities. As Linebaugh (2008) has suggested, the term "commoning" refers to the fluid, continuous and relational ways in which commons are produced. Moreover, commoning is a process of liberation from the exploitation of capitalist relations.

A wide range of these alternative ways of organization are elaborated upon in this volume. Özlem Ilyas researches the ways that commons-based coworking spaces in Istanbul, also rooted in social movements that mobilized during the Gezhi park protests in early 2010s, assist in building networks of solidarity among freelancers, while she questions the sustainability of such initiatives. Dimitra Gkitsa sees coworking spaces as contemporary modes of self-organization in the cultural sector in Southeast Europe and she argues that self-organized collective labor holds an affective power that can produce infrastructures and networks of solidarity in post-socialist countries. Both contributions discuss the formal and informal processes of exclusions and inclusions over resources, the negotiations and conflicts, the commons governmentality and the everyday management of common resources, the non-monetary exchanges in CWS, the sharing practices and the transformation of the "coworking communities" through commons arrangements.

On the other hand, Eeva Houtbeckers and Marita Korica explore the reasons for the organizational decline of a particular coworking space in Finland. Although that space was organized as a cooperative, it did not succeed in its goal, due to internal conflicts, opportunism, unpaid labor, and burnout. The lessons learnt from the decline of that community-led coworking space can be valuable for all those that aim to develop and sustain alternative coworking spaces. Finally, Dimitris Pettas and Vasilis Avdikos illustrate the impacts of the rise of digital nomads on coworking spaces in the city of Athens and discuss the turn of coworking into a more consumption-oriented service provision that caters to nomadic lifestyles. In turn, they argue that an alignment with the principles of the cooperative movement could reconstitute the coworking spaces with a new role in providing the terrain for emancipatory practices of labor solidarity.

Care and Labor in Coworking Practices

Coworking spaces are social spaces and constituted through myriad relationships inside, outside and across them. Research has identified diverse relational models that can develop in coworking spaces: from fleeting encounters to friendship and mutual support networks. However, most scholars emphasize network sociality (Wittel, 2001) as the dominant mode of sociality that sustains coworking (Bandinelli & Gandini, 2019; de Peuter et al., 2017). Similar to how freelance workers in creative industries are often depicted as "self-enterprising, self-reliant, self-interested and calculative agents who valorise care-free independence" (Alacovska & Bissonnette, 2021, p. 135), workers in coworking spaces have been described as individualized, competitive, resource-driven and instrumental (see, e.g., Bandinelli & Gandini, 2019; Grazian, 2020). For this collection, we are interested in specific types of social relationships that could foster social support and solidarity and that are based on recognizing mutual interdependencies. We assume that coworking spaces also might support practices of acting "other-wise instead of self-wise" (Alacovska, 2020, p. 735; Alacovska & Bissonnette, 2021; Lynch, 2007) where individual coworkers are not just self-centered and engaging in transactional relations but where a caring orientation is practiced and might transform social relationships and attitudes among coworkers.

Here, relational feminist theories of care can add interesting insights into coworking and its social relationships as care "begins with a social ontology of connection: foregrounding social relationships of mutuality and trust (rather than dependence)" as Lawson (2007, p. 3) notes. In this view, care is not seen "as a separate kind of relation, but as endemic to (potentially) all social relations that matter" (ibid). A care ethics also "suggests different ways of theorizing politics" (ibid) as it imagines actors as interdependent and not as autonomous subjects, and thus locates "care's political potential in its relationality" (Bartos, 2020, p. 264). Recent scholarship in geographies of care has extended the notions of care in cities and how care is practiced in and through specific spaces such as homes, day centers, shelters for homeless or community food shops (Power & Williams, 2020). We contend that coworking spaces can also be analyzed as sites of care, as places that enable and facilitate mutual support and where "care, as a capacity and a practice, is cultivated, shared and resourced on an egalitarian basis" (Collective, 2020, S. 6) because they are an important spatial context for the creation and maintenance of social relationships of independent workers. Yet care has not been used as an analytic framework, despite, for example, much evidence about the emotional labor of coworking spaces' hosts (Merkel, 2019) and the immaterial labor of coworkers in creating and maintaining the supportive communities within coworking spaces (Garrett et al., 2017). Even working from a coworking space is described by many as a form of self-care, as having positive benefits for their well-being because it creates boundaries for self-employed workers (Reissner et al., 2021). Therefore, coworking spaces as shared relational spaces could provide a social infrastructure for care practices, creating caring relationships and for fostering an ethics of care because caring practices are under-scored by elements of responsibility, attentiveness, responsiveness, and competence (Tronto, 1993). Thus, a care lens might help to direct our attention towards the construction of new forms of relationships, institutions, and actions that enhance mutuality and well-being in the context of coworking; to explore further the affective, emotional, and embodied dimensions of coworking and to extend our understanding of practices of care. It could give insights on the form and functions of care that emerge through coworking (i.e., informal care) and provides useful analytical insights into the practices and politics of interdependence and solidarity and the wider social role that coworking spaces might play in neighborhoods, cities, or regions (Care Collective, 2020).

The four chapters in this section help us understand how coworking spaces could be sites or infrastructures for the creation of meaningful, care-full social relationships, supporting empathy and an ethics recognizing mutual interdependencies as a foundation for social change. Janet Merkel's chapter discusses coworking spaces as social infrastructures of care through Joan Tronto's framework of five different phases of care. This analytical framework helps to see coworking spaces as sites of the everyday practices of situated, informal types of care and to think through the socio-political possibilities of coworking that emerge from these everyday practices. In a similar approach, Yoke Methorst's chapter proposes an embodied geographies of care approach and situates her empirical research in the specific context of Māori culture in New Zealand. She presents results from a cross-case analysis with 11 coworking hosts where practices of care articulate with notions of inclusivity, safety and care that are filtered through principles within Te Ao Māori (the Māori worldview). The following two chapters both provide a critical reading of forms of labor that sustain the social atmosphere and relationship building in coworking spaces and that remain mostly hidden from view in coworking as well as in coworking research. Antigoni Papageorgiou and Martha Michailidou discuss the invisible labor of community managers with empirical results from interviews in Athens and Berlin. They argue that the lack of a formal work description creates an informality around the work of community managers and perpetuates a condition of invisibility for their relational practices that are crucial for the maintenance of the community. However, this work is experienced as non-work by users, founders and even the people that carry out the work themselves. In the last chapter in this section, Nikos Gatsinos and Malte Höfner discuss the invisible labor of community members who spent a significant amount of their time in the coworking space for the maintenance of the community and the space. In using the concept of reproductive work for members' care work, the authors discuss the exhaustion and exploitation that some coworkers felt in community-led coworking spaces, and which let them to leave when they needed to grow professionally. However, they also show how former coworkers consider the sharing practices as crucial, and many miss this about working in a coworking space. Building on these insights, they argue that coworking can enable more than just feeding into the individualist neoliberal subjectivities it is often associated with.

Alternative Modes of Development Through Coworking and the Policy Responses

Coworking spaces are gradually emerging as distinct urban actors, interacting with the social, economic, cultural and political environments of the urban settings in which they are nested in, through a variety of both individual and collective practices, including the collaboration and the development of synergies with institutional and informal urban actors (regional and local governments, cultural institutions, actors of the market and the third sector, community and activist groups, residents' associations, social movements, etc.) and the engagement with local communities. The degree to which these relational incorporations of coworking spaces in the urban environment remain limited to individual practices and behaviors or give birth to transformative dynamics, leading to the overall diffusion of a collaborative culture on the neighborhood and city level, as well as the factors that enable or obstruct these developments remain vastly under-investigated. The same applies concerning the incorporation of coworking spaces in networks and governance arrangements that either further establish and reproduce mainstream developmental agendas and pathways or promote and materialize alternative modes of development. Moreover, the ways coworking spaces are incorporated in existing policy frameworks is also of crucial importance, along with the roles and associations allocated to such spaces in relevant regional and urban strategies.

In this vein, coworking spaces can undertake different roles and partake in diverse networks and governance arrangements. From their entanglement with the tourism sector or the culture and creative industries (CCI), to their emergence as part of the cooperative or social and solidarity economy (SSE) landscapes, coworking spaces undertake a wide range of characteristics, promoting different developmental visions, logics, and priorities. This diversity is also reflected upon their increasing, yet ambiguous incorporation in multi-leveled policies and strategies. Nevertheless, under the predominance of neoliberal regional and urban agendas, only a specific fragment of coworking spaces is commonly participating in institutional governance partnerships and targeted in developmental strategies, namely the one that can be aligned with mostly growth-driven understandings of development. On the other hand, more bottom-up and community-led coworking spaces also participate in networks and synergies that are built around societal goals and target to tackle local issues and challenges. The diverse ways in which coworking spaces emerge as urban actors, their intersections with networks and policies/ strategies are of key interest and their exploration is of significant academic and policy importance.

As such, in this volume, Alessandro Coppola explores the Mares de Madrid initiative that was established by the city of Madrid on the legacy of a variety of grassroots practices of the past and as a response to neoliberal urban models and austerity policies. The Mares di Madrid intended to create neighborhood-based and labor-intensive economies by opening new public collaborative facilities where new social and solidarity economy entities could be incubated. The chapter questions the initiative's transformative potential and discusses the ways that such innovative initiatives should be institutionalized through. Illaria Mariotti, Ivana Pais, and Francesca Ciccarelli discuss another public initiative the Presidi di Comunità that was established in small villages of the Southern Italy during the pandemic, and aimed to stimulate the local creative ecosystem and establish a fruitful relationship between remote workers and local communities, while also aiming at the development of local communities that have been particularly impoverished over the decades, in socio-economic terms and in terms of human capital due to brain drain.

On the other hand, Colm Stockdale and Lorenzo Marmo look at the ways coworking spaces and hubs engage with social innovation processes on the local level. Through the lens of assemblage thinking, they explore the social innovation dynamics that unfold around two social spaces in Athens, investigating how coworking spaces address social needs, improve social relations and support socio-political transformation in the neighborhoods they operate. In the same line, Muyi Chou and Salizan Takisvilainan research the role and function of a coworking space in an indigenous community in eastern Taiwan, through the lens of Diverse Economies. They analyze the ways that a university team builds relationships with community partners through the establishment of a multifunctional coworking space and how that initiative assists the community in reframing culture and publishing by assembling a diverse economy. Finally, Danai Liodaki sees collaborative workspaces as spaces of possibility for the transformation of local development paths. Researching the practices of a hub in a Greek peripheral town, she analyzes the ways that the hub aims to mobilize local communities to collectively question some deep-rooted perceptions for local development and peripherality and bring social change.

Emancipatory and Transformative Potentials of Coworking

Whether coworking constitutes a resource for postcapitalist (transformative) practice or rather substantiates the reproduction of exploitative social relations as "coworking spaces respond to and legitimize work precarization" (Fast, 2022, p. 1) is an important question underlying this collection. For now, there is little effort in the research field to understand the emancipatory or transformative social and political potentials that might emerge through coworking as a socio-spatial practice (Gandini & Cossu, 2021; Smith, 2020; Vidaillet & Bousalham, 2020). If emancipatory transformative potentials of coworking are discussed, cautionary views prevail (de Peuter et al., 2017; Gandini, 2015), highlighting the ambivalence and contradictions of coworking spaces both as spaces of alternative economic practices *and* as integrated into market economies (Richardson, 2017). Many scholars argue that coworking further entrenches the neoliberal entrepreneurial subjectivity (Bröckling, 2014) into contemporary modes of work (Grazian, 2020; Resch et al., 2021) with coworkers seen as "these atomized entrepreneurial subjects of neoliberalism" (Gandini, 2015, p. 202). In this view, coworking is rather seen as a function of precarity than offering means to manage it (Waters-Lynch & Duff, 2019, p. 3). There is some discussion in the coworking literature, where some scholars are "cautiously optimistic" about the emerging micro practices and their alterity. Argumentation range from "resilient spaces" (Gandini & Cossu, 2021), "post-capitalist practices" (Smith, 2020), "labour commons" (Avdikos & Pettas, 2021) to coworking spaces as "articulating economic diversity" for a "post capitalocentric economy" (Vidaillet & Bousalham, 2020, p. 1). Yet the debate lacks an empirical grounding and examples tend to be very limited. So, for this collection we were interested in theoretical, methodological, and empirical discussions that address the emancipatory and transformative potentials of coworking spaces. Several chapters in this volume address the emancipatory and transformative potentials of coworking spaces and point to the conditions where coworking spaces can be seen as sites of emancipatory practices.

How do we conceptualize transformative potentials? Here Eric Olin Wright's ideal types of transformative strategies for social emancipation which he discusses in *Envisioning Real Utopias* (Wright, 2010) are helpful to think ahead. Wright distinguished three ideal types of transformative strategies: The first ideal type specifies *symbiotic transformations.* Those are processes which address social issues and enhance possibilities for emancipation but without challenging capitalist institutions directly. They are mainly driven by social movements and work through pressure from below to compromises and negotiations. The second type of *interstitial transformations* involves strategies that build alternative forms of social organization in the "niches and margins of capitalist society" (Wright, 2010, p. 303) and, thus, provide important examples of alternative social arrangements. The third type specifies *ruptural transformations* that confront capitalist institutions and seek to create "new institutions of social empowerment through a sharp break with existing institutions and social structures." (ibd.) For now, research suggests that coworking spaces might support symbiotic and interstitial transformations. On the one hand, coworking spaces can perform as sites of expanding social power and as safety nets against labor precariousness, through different coalitions between social forces and labor, very often in collaboration with state and market actors (symbiotic transformation). On the other hand, there are signs that coworking spaces assemble a range of actors that mainly aim to operate and build alternatives and counter-hegemonic practices outside the state's spectrum and the capitalist market (interstitial transformation). The volume provides empirical data through case studies for both kinds of the transformative potential of coworking, looking not only to successful stories but also to cases that have not gone so well, and discuss the reasons for this. Moreover, the chapters that follow also provide the theoretical tools to analyze the ways that coworking spaces can play a role in the future as the progressive mediums of social and economic emancipatory transformations.

References

Alacovska, A. (2020). From passion to compassion: A caring inquiry into creative work as socially engaged art. *Sociology, 54*(4), 727–744. https://doi.org/10.1177/003803852090471

Alacovska, A., & Bissonnette, J. (2021). Care-ful work: An ethics of care approach to contingent labour in the creative industries. *Journal of Business Ethics, 169*(1), 135–151. https://doi.org/10.1007/s10551-019-04316-3

Avdikos, V., & Pettas, D. (2021). The new topologies of collaborative workspace assemblages between the market and the commons. *Geoforum, 121*, 44–52. https://doi.org/10.1016/j.geoforum.2021.02.012

Bacevice, P. A., & Spreitzer, G. M. (2022). 'It's like, instant respect': Coworking spaces as identity anchoring environments in the new economy. *New Technology, Work and Employment*, OnlineFirst 19 July 2022. https://doi.org/10.1111/ntwe.12254.

Bähr, U., Biemann, J., Lietzau, J., & Hentschel, P. (2021). *Rural Coworking. People, Models, Trends*. https://www.bertelsmann-stiftung.de/fileadmin/files/user_upload/BS_Coworking_im_laendlichen_Raum_ENG_v04_RZ_Einzelseiten.pdf

Bandinelli, C., & Gandini, A. (2019). Hubs vs networks in the creative economy: Towards a 'collaborative individualism'. In R. Gill, A. Pratt, & T. Virani (Eds.), *Creative hubs in question. Dynamics of virtual work* (pp. 89–110). Palgrave Macmillan.

Bartos, A. E. (2020). Relational spaces and relational care: Campus sexual violence, intimate geopolitics and topological polis. *Area, 52*(2), 261–268. https://doi.org/10.1111/area.12449

Boeri, T., Giupponi, G., Krueger, A. B., & Machin, S. (2020). Solo self-employment and alternative work arrangements. A cross-country perspective on the changing composition of jobs. *The Journal of Economic Perspectives, 34*(1), 170–195. https://www.jstor.org/stable/26873534

Bouncken, R. B., Aslam, M. M., & Qiu, Y. (2021). Coworking spaces: Understanding, using, and managing sociomateriality. *Business Horizons, 64*(1), 119–130. https://doi.org/10.1016/j.bushor.2020.09.010

Boutillier, S., Capdevila, I., Dupont, L., & Morel, L. (2020). Collaborative spaces promoting creativity and innovation. *Journal of Innovation Economics & Management, 31*(1), 1–9. https://doi.org/10.3917/jie.031.0001

Brakel-Ahmed, F., Cnossen, B., & Schlegelmilch, J. (2020). Discovering workscapes: An investigation of collective workspaces. In F. Brakel-Ahmed, B. Cnossen, & J. Schlegelmilch (Eds.), *Collaborative spaces at work* (pp. 37–49). Routledge.

Bröckling, U. (2014). *The entrepreneurial self: Fabricating a new type of subject*. Sage.

Butcher, T. (2018). Learning everyday entrepreneurial practices through coworking. *Management Learning, 49*(3), 327–345. https://doi.org/10.1177/1350507618757088

Capdevila, I. (2021). Spatial processes of translation and how coworking diffused from urban to rural environments: The case of Cowocat in Catalonia, Spain. In B. Hracs, T. Brydges, T. Haisch, A. Hauge, J. Jansson, & J. Sjöholm (Eds.), *Culture, creativity and economy* (pp. 95–108). Routledge.

Care Collective. (2020). *The care manifesto: The politics of interdependence*. Verso.

Clifton, N., & Reuschke, D. (2022). The diverse coworking landscape and implications for commercial real estate provision: Lessons from individual preferences and practice. *Journal of Property Investment & Finance, 40*(5), 508–523. https://doi.org/10.1108/JPIF-12-2021-0109

de Peuter, G., Cohen, N. S., & Saraco, F. (2017). The ambivalence of coworking: On the politics of an emerging work practice. *European Journal of Cultural Studies, 20*(6), 687–706. https://doi.org/10.1177/1367549417732997

de Peuter, G., Oakley, K., & Trusolino, M. (2022). The pandemic politics of cultural work: Collective responses to the COVID-19 crisis. *International Journal of Cultural Policy*, 1–16. https://doi.org/10.1080/10286632.2022.2064459

England, L., Iruobe, E., & Comunian, R. (2021). Coworking, gender and development: The case of tribe XX lab. In B. Hracs, R. Comunian, & L. England (Eds.), *Developing creative economies in Africa* (pp. 77–93). Routledge.

Fast, K. (2022). Who has the right to the coworking space? Reframing platformed workspaces as elite territory in the geomedia city. Space and Culture, 0(0). https://doi.org/10.1177/12063312221090429.

Felstead, A., Jewson, N., & Walters, S. (2005). *Changing places of work*. Palgrave Macmillan.

Gandini, A. (2015). The rise of coworking spaces: A literature review. *Ephemera: Theory & Politics in Organization, 15*(1), 193–205.

Gandini, A., & Cossu, A. (2021). The third wave of coworking: 'Neo-corporate' model versus 'resilient' practice. *European Journal of Cultural Studies, 24*(2), 430–447. https://doi.org/10.1177/1367549419886060

Gandini, A., Bandinelli, C., & Cossu, A. (2017). Collaborating, competing, co-working, coalescing: Artists, freelancers and social entrepreneurs as the 'new subjects' of the creative economy. In A. Gandini & J. Graham (Eds.), *Collaborative production in the creative industries* (pp. 15–32). University of Westminster Press. https://doi.org/10.16997/book4.b

Garrett, L. E., Spreitzer, G. M., & Bacevice, P. A. (2017). Co-constructing a sense of community at work: The emergence of community in coworking spaces. *Organization Studies, 38*(6), 821–842. https://doi.org/10.1177/0170840616685354

Grazian, D. (2020). Thank god it's monday: Manhattan coworking spaces in the new economy. *Theory and Society, 45*, 991–1019. https://doi.org/10.1007/s11186-019-09360-6

Harris, R. (2021). *London's global office economy: From clerical factory to digital hub*. Routledge.

Herod, A. (1997). From a geography of labor to a labor geography: Labor's spatial fix and the geography of capitalism. *Antipode, 29*(1), 1–31. https://doi.org/10.1111/1467-8330.00033

Howell, T. (2022). Coworking spaces: An overview and research agenda. *Research Policy, 51*(2), 104447. https://doi.org/10.1016/j.respol.2021.104447

Huron, A. (2015). Working with strangers in saturated space: Reclaiming and maintaining the urban commons. *Antipode, 47*(4), 963–979. https://doi.org/10.1111/anti.12141

Kalleberg, A. L. (2018). *Precarious lives: Job insecurity and Well-being in rich democracies*. Polity Press.

Krasilnikova, N., & Levin-Keitel, M. (2022). Mobile Arbeit aus sozioräumlicher perspektive: Co-Working Spaces als Chance für suburbane und ländliche Räume? *Raumforschung und Raumordnung|Spatial Research and Planning, 80*(3), 360–372. https://doi.org/10.14512/rur.115

Kraus, S., Bouncken, R. B., Görmar, L., González-Serrano, M. H., & Calabuig, F. (2022, January 1). Coworking spaces and makerspaces: Mapping the state of research. *Journal of Innovation & Knowledge, 7*(1), 100161. https://doi.org/10.1016/j.jik.2022.100161

Lawson, V. (2007). Geographies of care and responsibility. *Annals of the Association of American Geographers, 97*(1), 1–11. https://doi.org/10.1111/j.1467-8306.2007.00520.x

Linebaugh, P. (2008). *The magna carta manifesto: Liberties and commons for all*. University of California Press, Berkeley, CA.

Lorne, C. (2020). The limits to openness: Co-working, design and social innovation in the neoliberal city. *Environment and Planning A: Economy and Space, 52*(45), 747–765. https://doi.org/10.1177/0308518x19876941

Luo, Y., & Chan, R. C. K. (2020). Production of coworking spaces: Evidence from Shenzhen, China. *Geoforum, 110*, 97–105. https://doi.org/10.1016/j.geoforum.2020.01.008

Lynch, K. (2007). Love labour as a distinct and non-commodifiable form of care labour. *The Sociological Review, 55*(3), 550–570. https://doi.org/10.1111/j.1467-954X.2007.00714.x

Mariotti, I., Di Vita, S., & Akhavan, M. (Eds.). (2021). *New workplaces–Location patterns, urban effects and development trajectories: A worldwide investigation*. Springer.

Mariotti, I., Di Marino, M., & Bednář, P. (Eds.). (2022). *The COVID-19 pandemic and the future of working spaces*. Routledge.

McNeill, D. (2021). Urban geography 1: 'Big tech' and the reshaping of urban space. *Progress in Human Geography, 45*(5), 1311–1319. https://doi.org/10.1177/03091325211021945

Merkel, J. (2019). Curating strangers. In R. Gill, A. Pratt, & T. Virani (Eds.), *Creative hubs in question. Dynamics of virtual work* (pp. 51–68). Palgrave Macmillan.

Merkel, J. (2022). Coworking spaces as destinations and new stakeholders in hospitality ecosystems. In N. Olbrich, J. Philipp, H. Thees, & H. Pechlaner (Eds.), *Towards an ecosystem of hospitality–Location: City: Destination* (pp. 140–147). Graffeg Limited.

Moisio, S., & Rossi, U. (2020). The start-up state: Governing urbanised capitalism. *Environment and Planning A: Economy and Space, 52*(3), 532–552. https://doi.org/10.1177/0308518x19879168

Montgomery, J. (2007). Creative industry business incubators and managed workspaces: A review of best practice. *Planning, Practice & Research, 22*(4), 601–617. https://doi.org/10.1080/02697450701770126

Murgia, A., & Pulignano, V. (2021). Neither precarious nor entrepreneur: The subjective experience of hybrid self-employed workers. *Economic and Industrial Democracy, 42*(4), 1351–1377. https://doi.org/10.1177/0143831x19873966

Power, E. R., & Williams, M. J. (2020). Cities of care: A platform for urban geographical care research. *Geography Compass, 14*(1), e12474. https://doi.org/10.1111/gec3.12474

Pulignano, V., Domecka, M., Muszyński, K., Vermeerbergen, L., & Riemann, M.-L. (2021). *Creative labour in the era of COVID-19: The case of freelancers* (ETUI research paper-working paper). https://doi.org/10.2139/ssrn.3832553.

Rådman, E., Johansson, E., Bosch-Sijtsema, P., & Raharjo, H. (2022). In search of member needs in coworking spaces. *Review of Managerial Science*, 1–27. https://doi.org/10.1007/s11846-022-00546-4

Resch, B., Hoyer, P., & Steyaert, C. (2021). Affective control in new collaborative work: Communal fantasies of purpose, growth and belonging. *Organization Studies, 42*(5), 787–809. https://doi.org/10.1177/0170840620941616

Reuschke, D., Clifton, N., & Fisher, M. (2021). Coworking in homes – Mitigating the tensions of the freelance economy. *Geoforum, 119*, 122–132. https://doi.org/10.1016/j.geoforum.2021.01.005

Richardson, L. (2015). Performing the sharing economy. *Geoforum, 67*, 121–129. https://doi.org/10.1016/j.geoforum.2015.11.004

Schmidt, S., Brinks, V., & Brinkhoff, S. (2014). Innovation and creativity labs in Berlin organizing temporary spatial configurations for innovations. *Zeitschrift fuer Wirtschaftsgeographie, 58*(4), 232–247. https://doi.org/10.1515/zfw.2014.0016

Shearmur, R., Ananian, P., Lachapelle, U., Parra-Lokhorst, M., Paulhiac, F., Tremblay, D.-G., & Wycliffe-Jones, A. (2021). Towards a post-COVID geography of economic activity: Using probability spaces to decipher Montreal's changing workscapes. *Urban Studies, 59*(10). https://doi.org/10.1177/00420980211022895

Smith, T. S. (2020). 'Stand back and watch us': Post-capitalist practices in the maker movement. *Environment and Planning A: Economy and Space, 52*(3), 593–610. https://doi.org/10.1177/0308518X19882731

Spinuzzi, C. (2012). Working alone together: Coworking as emergent collaborative activity. *Journal of Business and Technical Communication, 26*(4), 399–441. https://doi.org/10.1177/1050651912444070

Spinuzzi, C., Bodrožić, Z., Scaratti, G., & Ivaldi, S. (2019). "Coworking is about community": But what is "community" in coworking? *Journal of Business and Technical Communication, 33*(2), 112–140. https://doi.org/10.1177/1050651918816З

Stevens, L., & Shearmur, R. G. (2020). The end of location theory? Some implications of micro-work, work trajectories and gig-work for conceptualizing the urban space economy. *Geoforum, 111*, 155–164. https://doi.org/10.1016/j.geoforum.2020.02.010

Tintiangko, J., & Soriano, C. R. (2020). Coworking spaces in the global south: Local articulations and imaginaries. *Journal of Urban Technology, 27*(1), 67–85. https://doi.org/10.1080/10630732.2019.1696144

Vallas, S., & Schor, J. B. (2020). What do platforms do? Understanding the gig economy. *Annual Review of Sociology, 46*, 273–294. https://doi.org/10.1146/annurev-soc-121919-054857

Vidaillet, B., & Bousalham, Y. (2020). Coworking spaces as places where economic diversity can be articulated: Towards a theory of syntopia. *Organization, 27*(1), 60–87. https://doi.org/10.1177/1350508418794003

Waters-Lynch, J., & Duff, C. (2019). The affective commons of coworking. *Human Relations, 74*, 383–404. https://doi.org/10.1177/0018726719894633

Wittel, A. (2001). Toward a network sociality. *Theory, Culture & Society, 18*(6), 51–76. https://doi.org/10.1177/026327601018006003

Worth, N., & Karaagac, E. A. (2021). Accounting for absences and ambiguities in the freelancing labour relation. *Tijdschrift voor economische en sociale geografie*. https://doi.org/10.1111/tesg.12491.

Wright, E. O. (2010). *Envisioning real utopias*. Verso.

Part I
Coworking and the Politics of Self-Organization and Community Organization in the City

A Place in the World: A Mapping of Coworking as a Site/Practice of Commoning and Community in Istanbul

Özlem İlyas

Introduction

The rise of coworking spaces has been associated with the increasing flexibilization of labor markets, the adoption of austerity policies in the post-2008 period, the concomitant informalization of labor (Merkel, 2019), and the commodification of material and immaterial means of production (Jakonen et al., 2017) in knowledge-based and cultural industries. While some studies point to the tension-ridden co-existence of the possibility of co-optation of such spaces by the neoliberal market logic and their potential to create alternatives to it (Merkel, 2019; Gandini, 2015, Bandinelli & Gandini, 2019), few studies seem to try to distinguish between different types of coworking spaces to delineate a distinction between these two possibilities and point to the conditions under which they play out (Avdikos & Iliopoulou, 2019; Gandini & Cossu, 2021). This study is aimed to contribute to the latter endeavor by providing a comparative analysis of two private, one solidarity-based and one public coworking space in Istanbul.

The prevalence of mostly empiricist approaches in critical accounts of coworking strives to determine the potential benefits and/or disadvantages of the coworking phenomenon for precarious workers (Gandini, 2015). A few studies analyze the conditions of existence of such spaces in relation to the withdrawal of the state under neoliberalism (Merkel, 2019); however, the question of how the differing types of coworking spaces help undermine or reproduce the exploitative relations of production is often evaded. This leads to a difficulty in understanding the conditions of existence of such potentials, turning research into a form of cost-benefit analysis of coworking for precarious workers. In this study, it is argued that class and non-class processes (Resnick & Wolff, 1987) productive of and produced in such spaces

Ö. İlyas (✉)
University of Warwick, Department of Sociology, Coventry, United Kingdom
e-mail: Ozlem.Ilyas-Savk@warwick.ac.uk

J. Merkel et al. (eds.), *Coworking Spaces*, https://doi.org/10.1007/978-3-031-42268-3_2

needs to be analyzed to provide an understanding of the distinctive properties of capitalist-run, commons-based and public spaces in which coworking takes place. The focus of the present study is on the experiences of freelancers in such spaces, as coworking is often considered as a solution to the aggravating conditions of precarity and informality.

Research Sites and Methodology

The multi-sited ethnography presented here is based on a larger research conducted on freelancing in Turkey. It involved long-term activist-participant observation, in-depth interviews with more than 20 freelancers from different industries and several focus-group research aimed at establishing a freelancers' network.[1] The research involved a specific attention to the different sites in which freelancers work, and it is this aspect of their experience which informs the analysis presented here. The first sites under scrutiny are two privately-owned and operated coworking spaces located in central districts of Istanbul. They were visited multiple times in 2017 and their managers and four member users were interviewed. The second site to be analyzed is an urban common called *Dünyada Mekân* [A place in the world] which was founded by white-collar workers, freelancers and unemployed individuals in the aftermath of the Gezi Uprising in Istanbul,[2] and operated between the years of 2015–2019. Used as a co-working space by freelancers free of charge in daytime and as a meeting venue by various grassroots organizations in the evening, it enabled encounters that led to the foundation of a network of solidarity called *Ofissizler* [the Officeless] among freelancers. The findings for that site are based on the author's participation in the constitution and operation of that space and in-depth interviews conducted with six other founding members between 2016 and 2017. The author also participated in several focus-group workshops which were held in *Dünyada Mekân* and led to the foundation of the freelancers' network mentioned before.

Finally, a public coworking space called IDEA is analyzed in comparison to the aforementioned ones, towards providing a tentative map of the different relationalities imagined and enacted in spaces of different "ethe". It is run by the municipality of Kadıköy in Istanbul and is a two-floor building, part of which can be used by citizens free of charge. Sometimes, it also provided a meeting venue for grassroots organizations such as *Ofissizler*, and the author participated in the organization of

[1] It was conducted for the M.A. dissertation thesis submitted to Boğaziçi University, Department of Sociology in 2019.

[2] Gezi uprisings emerged in 2013 when the urban activists' attempts to prevent the demolishing of a park at the central district of Taksim in Istanbul which was to be turned into a shopping mall drew the attention of the politicians and citizens at large. The use of humor, the preference for horizontal, bottom-up way of organizing and the carnivalesque ways of protesting were some of the emphasized qualities of the resistance. For an example, please see the edited volume by David and Toktamış (2015).

such meetings and held informal conversations about its management with its managers. An in-depth interview with its management had also been conducted when the space was first opened in 2017.

The sites in question are not imagined as bounded locations with unified cultures (Gupta & Ferguson, 1997), but are approached as sites of contesting and contradicting ethe. Pointing to contestations in that regard is aimed to re-politicize the representation of socio-economic relations as politics is taken to be related to the desire and capacity to deal with conflict and ambiguities (Byrne & Healy, 2006). Also, rather than providing a "realistic" representation of different coworking spaces, the aim of the analysis is to problematize the differing representations of economy their discourses imply. J.K. Gibson-Graham's (2006) work is inspirational in that regard since it emphasizes the need to both criticize the "capitalo-centric" representations of the economy which effaces economic diversity and to problematize the affective grip the reigning discourses have on political subjectivity. For the latter aim, the present study employs a psychoanalytically inflected discourse pointing to the role of disempowering affects in depoliticizing the socio-economic relations in the first two sites under scrutiny.

Imagine It Like an Ecosystem!: Neoliberal Discourse on Coworking

The impact of the discourse on coworking in spaces operated by capitalist firms needs to be preceded by an account of the experience of freelancers for whom such spaces are often seen as palliative solutions against precarity (Bandinelli & Gandini, 2019; Blagoev et al., 2019). During the in-depth interviews, freelancers often reported experiences of mobbing, unpaid overtime, low wages, long working hours, traffic jam, being laid out and their impact on their mental and physical health in their office jobs as among the reasons they started freelancing. In some industries such as publishing, translation and journalism, freelancing has become the "new normal" with steady jobs diminishing dramatically.[3] So, freelancers often have either traumatic experiences in workplaces or are excluded from such spaces, which leads to their "responsibilization" for the organization of work and non-work realms of life (İlyas, 2022). Being "officeless" also means loss of spatio-temporal commonality with co-workers, bringing with it an aggravated sense of isolation that leads to the individualization of the work process.

Coworking spaces are claimed to offer solutions to confront that sense of isolation, enabling freelancers to have community, social interaction, discipline, productivity, and routines (Brown, 2017; Blagoev et al., 2019), while the extent to which such promises are achieved are also questioned (Brown, 2017; Spinuzzi et al., 2019;

[3] An international report conducted in six countries in 2016 by Ehrensberger-Dow et al. claim 78% of translators worked as freelancers.

Gandini, 2015), especially in non-Western contexts (Tintiangko & Soriano, 2020). There is also a large body of research analyzing the management of such spaces, with the aim to establish typologies of spaces and their managers (Kojo & Nenonen, 2016; Ivaldi et al., 2018). However, what is missing in such accounts is the distinctive way in which labor is organized in the firms operating those spaces, the class processes their members go through in time, and the social and political implications of the discourse they employ in addressing target groups. This tendency unfortunately aligns with the "naturalistic" discourse of capitalist-run coworking spaces which covers over class diversity and antagonism (Gibson-Graham, 1996; Madra & Özselçuk, 2010).

Before elaborating on this discourse further, elucidating the class definition employed here is necessary. Class is hereby used as an adjective referring to the different organization of the processes of the production, appropriation, and distribution of surplus (Resnick & Wolff, 1987). The minimum definition of exploitative class processes is that they involve the exclusion of the producers from the moment of appropriation. Madra and Özselçuk (2010) draw on that class-process definition to formulate an ontological conception of "class antagonism" from a psychoanalytical viewpoint, claiming that the working of surplus *jouissance* would render any institution and subject's relation to it unstable, including the different ways in which class processes are organized. So, all class processes including the non-capitalist ones are to be open to constant challenging, negotiation, and conflict.

Coworking as Mutualism

Two different modes of doing away with that antagonism seem to be at stake in the capitalist-run coworking spaces visited. The first involves its disavowal through what could be called a "mutualist" discourse, promoting a conflict-averse depoliticizing perspective on sociality. It is a chain of coworking hubs located currently in seven central business locations in Istanbul, and the visit was paid to the oldest ones of those locations in Sanayi and Levent, where the manager of the space was interviewed in 2017. The space was visited again so that the author would participate in an entrepreneurship event, and two users were interviewed thereafter. The discourse of the manager pointed to a conflict-free approach to the "ecosystem" which is highlighted as the distinctive quality of the space in contrast to "service-providing" coworking spaces. The space in Sanayi, which was closed down later, was small and claimed to cater to the needs of creatives, freelancers and it was suggested that large companies are not accepted in fear of disrupting that ecosystem. On the other hand, in Levent, which is a much larger space, it was emphasized that the reason why people prefer their space was that the largest five investors in Turkey were a part of it. Different agents such as investors, creative agencies, corporations, service provider professionals, start-ups and freelancers were claimed to be always open to collaboration and benefit from each other. The manager stated "*Imagine it like the ecosystem in nature, all living beings, all those who live in it feed on each other.*"

The manager's following emphasis was on the autonomy of this ecosystem in relation to the economic crises and transformations in Turkey, thanks to the connections of the agents with each other and the global markets in general.[4] So, the mutualist discourse disavows not only class diversity and antagonism, but also social interdependence, by replacing crisis-prone society with harmonious "community." It contradictorily emphasizes the importance of social relations while promising freedom from them in case they ends up being costly. This simultaneous acknowledgement and denial regarding the social alludes to the perverse structure in Lacanian psychoanalysis.[5] Octave Mannoni famously summarizes the split of consciousness in perversion as "I know very well, but just the same . . ." (quoted in Copjec, 2002, p. 221). If we apply that to the mutualist discourse predicated on a conception of freedom in sovereignty, we could rewrite it as follows: we know very well that the subject is lacking and we are dependent on each other as social beings, but just as the same we are encouraged to pretend to be sovereign subjects without lack, in control of our relations to others and reigning supreme over our spatio-temporal and laboring conditions.

Coworking as Curation

A similar inward-oriented and conflict-averse discourse was employed by the manager of another privately-owned coworking space located in Bomonti, which is a historical site in Istanbul. This time conflict is foreclosed through the "curation" of members in the community. The space attempted to have a balanced distribution of members in three different sectors defined as creatives, technicians, and connectors. The manager claimed to be very selective in their acceptance of new members, stating that they only accept 15–20% of the applicants. New applicants propose a project which needs to be found suitable and attractive to the managers of the community and in observance of its effect on the pre-existing constituents of the space. It was emphasized that members cannot simply focus on their work and must work on creating collaborations. Otherwise, the manager stated that they are kindly asked to leave the community. So, while the first space promises a consoling indifference to others in line with one's sovereign interests, this space seems to banish it in favor of

[4]This promise of access to global markets was also interestingly seen in Greece (Papageorgiou, 2016).

[5]The relation of perverse structure to sovereignty needs to be explained here. Bruce Fink argues that in perversion, the subject is alienated in the symbolic but not separated from the other as her desire is not named (2003, p. 48). The subject then experiences the anxiety of the possibility of being engulfed by it, which she manages by propping up the law or offering herself as the object *a* to make up for the lack of lack (p. 50). Copjec (2002) relates that assumption of sovereignty on the part of the subject through the disavowal of lack with foregoing of subjectivity itself, because it involves avoiding the very split that characterize the neurotic subject. In contrast to the neurotic who questions and challenges the other, the pervert has ready-made, self-formulated answers, strategies, rules and regulations.

enforced compatibility. The imaginary register here seems to cancel the possibility of conflict from the beginning, rather than disavowing it.

Contradictory Appeal of Sovereignty and (In)dispensability

This depoliticizing conception of community is accompanied by a discourse of freedom in sovereignty when it comes to freelancing, which involves another contradiction as to social interdependence. It is acknowledged that places and people are indispensable for networking, yet at the same time, they are considered dispensable for this sovereign subject who is claimed to be able to work wherever, with whomever, whenever she wants. Coworking spaces also seem to think of their independent members as dispensable, given that they focus on providing office space to larger companies in the post-COVID-19 period.[6]

An anecdote from one of their users interviewed could be explanatory of the ethico-political ramifications of such a commodification of community. She used the space as an aspiring start-upper trying to broaden her network in one of the coworking spaces mentioned above. She pointed to the difficulty of doing so without family support, mentioning her realization that such spaces were not targeting her in the first place. Her account of one of the breakfast events organized by the space for networking is illustrative of her shock in encountering community on sale:

> The weather was a bit cold; but anyways, we went in. It was written nowhere that it was charged, and we assumed that it was free of charge. Even Ayşe assumed that it was free; she had membership in the space. She had to pay too. We filled our plates; Elif and I had very little. They looked at it and charged us 100 liras. I felt riveted to the ground and blushed. I asked, "Is it really charged?"

So, the neoliberal discourse does not simply commodify material and immaterial means of production, but also strives to hide that commodification. It is this hiding that instigates hiding on the part of the subject, who feels astonished and ashamed for thinking that networking could involve relations of gift, rather than exchange. Shame is an affect relating to the unassimilability of desire to social ideals, involving a distancing of the subject from them through an impulse to hide (Copjec, 2002, p. 179). Here it points to the limit of disavowal of class inequalities and social interdependence promoted by the mutualist discourse on "community."

On the other hand, this conception of sociality has not only depoliticizing, but also disempowering effects on the subject. Finding work requires a lot of unremunerated labor for socializing. The sovereign is to assume responsibility for that, which leads her to feel guilt in cases of exclusion or burnout. Copjec (2006) explains that in guilt, the subject deprives herself of any privacy, judging herself by social

---

[6]This tendency is remarked by a recent report from Britain, for details please see: https://digit-research.org/blog_article/coworking_in_the_city_after_covid/ (accessed on 7 November 2022). It also came up in the follow-up interview I conducted with a start-up owner working in such spaces in 2021.

ideals to avoid "the responsibility of having to invent a future without the aid of rules or scripts" (p. 24). The belief in ideals is sustained by fantasies of transgression, imagining that if she could "go beyond" herself, she could be successful. That keeps the self-blaming subject responsible and sovereign at the same time.

Finally, the discourse of sovereignty contradictorily implies a disempowering chronic infliction of violence on the self, while any existence of negativity in the community is denied at the same time. This could contribute to a tendency to turn on oneself and become reluctant subjects avoiding conflict-prone social relations. This desire was shared by a freelancer who participated in one of the workshops organized by *Ofissizler*. He was in the advertising industry and suggested "not wanting to deal with people anymore" as among many people's reasons to go freelance. Considering that freelancers often have traumas relating to social relations in the workplace, the mutualist discourse could further aggravate their reluctance to bond through its fantasmatic representation of community as a conflict-free service undercut by a guilt-ridden affective economy.

A Sublimated Space of the Commons: *Dünyada Mekân*

Instead of relations of exchange of "equivalents" qua commodities, a practicc of commoning and gift exchange marked the experience of coworking in a space called *Dünyada Mekân* [A place in the world] in Istanbul. It was a self-funding coworking space founded by an initiative that emerged through a series of fora organized by white-collar workers, freelancers and the unemployed in the aftermath of the Gezi resistance and operated as a meeting venue for grassroots organizations during the evenings. The initiative itself also hosted events on issues regarding white-collar workers and precarity such as indebtedness, competition, and private employment agencies. It defined itself as a space "without bosses or managers," illustrating the

Fig. 1 Table set for the opening of an artist's exhibition in 2016, published on the Facebook page of *Dünyada Mekân*

desire for autonomy on behalf of the workers outside the surveillance of the workplace (Fig. 1).[7]

The space was opened and managed in rotation by initiative members, who were also responsible for various maintenance tasks. Besides workshops to share knowledge in such fields as video-editing, translation and web design, the space hosted various workshops for "experience sharing" which led to the emergence of *Ofissizler*. The first of those workshops was organized among freelancers in the publishing industry in 2015, and involved a discussion of various issues such as informality, blurring of the boundaries between work and non-work, self-management, income security, job description, control over work, sense of isolation and pressure to socialize for networking. Then, two more workshops were organized by freelancers from various industries in 2016. In the first one, freelancers discussed various issues and then decided to focus on networking for the following workshop. *Ofissizler* was then founded as a network of solidarity for freelancers following two other workshops in 2018 in *Dünyada Mekân*. An important point made concerned the relation between the lack of recognition of freelancing as a distinct form of employment and the difficulty of formulating demands and solutions. To overcome this nonrecognition instigated the emergence of the network in *Dünyada Mekân*. It enabled freelancers to see the issues they experience often in isolation as social problems related to working conditions at large, hence worked to overcome feelings of guilt that prevent collective action.

Besides formulating rights-based demands for freelancers, the space also hosted desires and actions in favor of class transformation for freelancers. It hosted a collective of consumers called *Dürtük* [Resisting Producers and Consumers Collective], providing groceries from the small producers in the peripheries of Istanbul, and a collective of three freelancers named *Bağımsız Atölye* [Independent Workshop] who started a design and IT collective to produce and share its surplus together. The latter experience enabled a partial shift from independent and capitalist class processes to communal and independent ones for their members, while the first collective was mostly supporting the continuation of small producers' farming.

The kind of sociality imagined and enacted in *Dünyada Mekân* should have had a supportive role in hosting and flourishing of these political subjectifications. The space itself did not have a pre-determined agenda or political goals, hence could be approached as an urban commons open to use and experimentation by white-collar workers and freelancers. It hosted artist talks, mostly on topics related to labor and grassroots politics, film screenings, swap markets, report presentations on labor strikes and work experiences of trans women, reading gatherings, interviews on indebtedness, precarity and the ecological crisis. While social interdependence was acknowledged in relations of sharing experience, knowledge and work, class transformation found expression in experimentations with non-capitalist ways of production and consumption.

[7] For details, please see: https://dunyadamekan.wordpress.com/bilgi/ (retrieved on 17 October 2017).

In other words, *Dünyada Mekân* was an inspiring space of encounter and sublimation (Madra & Özselçuk, 2015), enabling an affective disinvestment from existing ways of being and acting, trying to empower its participants to politicize what exists and imagine and enact alternatives to it. The community hence established was neither one of exchange, nor based on a "common being" in terms of communitarianism. Both mutualist rendering of entrepreneurial ethos and the communitarian conception of community end up being inward-oriented and exclusionary of those who lack the identities desired.

Challenges of Commons-Based Coworking

One of the challenges of sustaining such urban commons is that they run the risk of returning to communitarianism on the one hand, disintegration on the other for several reasons. As mentioned above, social link in commons is based mainly on gift relations rather than relations of exchange. Gift is often counterposed to exchange in being free from competitive market mechanism; however, it does not mean that gift excludes points of antagonism. Gift incurs debt, and ways to deal with that has even led to claims about the impossibility of pure gift (Derrida, 1995). On the other hand, reciprocity is one of the norms in gift (Osteen, 2010, p. 570), and could be argued to constitute a point of antagonism, with our social entanglement producing a sense of anxiety that could be difficult to deal with on the part of the subject. If not dealt with through some sort of institutional mechanisms, such practices themselves may turn the subject away from commoning practices. It could incur a desire to go back to pre-given ways of relating to others, which is something we observed among some of the visitors of *Dünyada Mekân*. Instead of dealing with others, it could seem more comfortable to pay the price of the coffee and mind one's business on a hectic workday.

Secondly, if commercialization of the social is at stake in coworking spaces operated by capitalist firms, communitarianism could be the risk that could undermine the operation of such initiative-based spaces as commons. It could turn into a closely knit community and produce a homogeneity that prevents the entrance of others to the commons. Another risk is that when the members of that community lose interest, it runs the risk of depreciation as interpersonal relations rather than impersonal structures support its functioning. Commoning, on the other hand, requires a different relation to the others and the community for both its sustenance and political force. It would require an avowal of differences and the concomitant conflicts as a precondition of their politicization. In that case, we need a post-fantasmatic relation to community, which would involve the subject receiving satisfaction from dealing with conflicts, antagonisms, and contingencies (Byrne & Healy, 2006). It would also require an acknowledgement of the non-existence of any positive grounds for sociality, which means that every social process and relation could be open to politics, rather than being reduced to relations of service and exchange.

Another important determinant contributing to the sustainability of such commons and communities is related to the precarious working conditions. Freelancers often have unstable sources of income, working hours and work/non-work relations, while organizing and sustaining a common space requires some stability. This issue has marked the experience of *Dünyada Mekân* as well, and it became difficult to sustain it for several material reasons. In the beginning, we rotated the task of opening the space so that it could be used by any freelancer who decides to stop by. However, in time, those who volunteered had to either take full-time jobs due to not being able to sustain freelancing, or some started to find the space uncomfortable. It was very cold in winter, there happened to be security issues related to our belongings, and cleaning turned into an issue. In addition, regular contributions to rent and utilities declined, and as the space started to be used less, its maintenance problems mounted. On top of that, in 2016, several bombings in different parts of Istanbul[8] one of which happened on Istiklal Street very close to *Dünyada Mekân*, led to a diminishing of the use of public spaces for some time in general and of our space in particular. Finally, the space was actually a historical building, being rented by a trustee appointed on behalf of a family living abroad. The trustee decided to increase the rent substantially, and there happened to be rumors that it was going to be used for an urban renewal project, similar to the transformation of several buildings in Taksim.[9] Together with the sustainability issues mentioned, this added to the final decision to close the space in 2019.

To conclude, *Dünyada Mekân* enabled the emergence of some creative tools of intervention in capitalo-centric discourses and practices such as experience sharing workshops, skills workshops aiming to redistribute skills among members, non-capitalist forms of producing and consuming and the establishment of a solidarity network among freelancers. On the other hand, the community thus created seemed to lack a common signifier to establish links among these tools (Gibson-Graham, 2006), or mechanisms to deal with the antagonism that ensues from class and non-class differences. This seems to have contributed to a loss of interest in commoning, which was already under threat due to the violence disrupting the public space and the pre-existing traumatic work relations of the precarious mentioned above.

[8]The one in Beyoğlu happened on 19 March 2016, killing four people. Besides the attacks in 2016, the attack on the Peace Meeting organized in Ankara in 2015 following the escalation of violence in the Kurdish region is also important for the public memory. The marking of public space by such violence led to the diminishing of political mobility to a large extent.

[9]The landmark event preceding the Gezi protests in Taksim, Istanbul was related to the demolition of a movie theatre to build a shopping mall. Urban activists organized to prevent this demolition and protest the commodification of public space in 2013, and their struggle was turned into a documentary and archive project that could be followed here: https://ozgurlesenseyirci.com/ (accessed on 13 March 2022). Gezi Parkı could not be demolished and replaced with a shopping mall; however, several sites on Istiklal and Taksim in general have undergone gentrification in the meantime. A recent case is the "renewal" of Narmanlı Han in 2018, which was a building of 187 years old, having hosted several artists and writers in the meantime. It was bought by a cosmetic company and turned into a shopping mall. For details, see https://m.bianet.org/bianet/kent/201280-beyoglunun-187-yillik-yeni-avm-si-narmanli (accessed on 13 March 2022).

Coworking in the Public: As a Right to the City?

Could the "right to the city" be such a signifier, enabling an investigation on the sustainability of the commons and further socialization of work and non-work realms of our lives? The ambivalence of commoning in relation to the neoliberal restructuring of public space and production relations has been noted so far (Enright & Rossi, 2018). Such claims could be criticized for having a capitalo-centric view of social actors and practices[10]; however, the question remains. Could we define and demand coworking as a right to the city by/for the precarious workers (Harvey, 2008)?

Following the opening of *Dünyada Mekân* in Beyoğlu, the municipality of Kadıköy opened a public coworking space in its central district which seems to have a large population of freelancers.[11] The space has two floors, with the ground floor being open to use free of charge and the second floor rented to freelancers and start-uppers for a cheaper amount than the privately owned coworking spaces. *Ofissizler* has also used the venue for one of its workshops where it presented a survey result about the contract use of freelancers and informed them about their legal rights. It was attended by legal experts who presented about the kinds of contracts freelancers could use according to Labour Code in Turkey, as well as the legal courses of action they could follow in cases on non-payment (Fig. 2).

The use of public spaces for coworking has been important for *Ofissizler* since the closing of *Dünyada Mekân*. It used an artist's space in Kadıköy every Thursday and another space belonging to an environmental organization in Beyoğlu every Wednesday to cowork until the COVID-19 pandemic.[12] Thereafter, it started using a pre-existing chatting platform for freelancers to cowork online. It went well for the first year; however, the second year in the pandemic led to a loss of interaction, which diminished the labor put in the maintenance of both the solidarity network and the online platform where it often met. The coworking days are now claimed to

[10] Gibson-Graham et al. (2000) propose that capitalo-centric view represents the social space as singularly capitalist, thinking of non-capitalist practices (such as the ones encountered in commoning) as either transitory or supportive of the capitalist processes.

[11] We do not have any quantitative data about the number or distribution of freelancers in Istanbul. Yet as *Ofissizler*, we went to that district to do street interviews and asked people what they know about freelancing. Surprisingly, most of the people we stopped to ask questions turned out to be freelancers. We then published an edited video based on the interviews on the *Ofissizler*'s website: https://ofissizler.org/neymis-bu-frilens/ (accessed on 10 March 2022).

[12] The first space called *arthere* was founded in 2014 with the aim of increasing collaboration between Syrian and Turkish artists, while also providing residency to the international community of artists. It also had a coworking space/cafe which Ofissizler used in collaboration with the space. For more information about the space, see https://www.arthereistanbul.com/home (accessed on 13 March 2022). The second space belonged to an environmental organization called *Kuzey Ormanları Savunması* (Defense of the Northern Forests) which was founded in 2013 to defend the forests and water resources located in the north of the Marmara region, spanning cities such as Edirne, Kırklareli, Tekirdağ, İstanbul, Kocaeli, Yalova, Sakarya, Düzce Bursa, Balıkesir and Çanakkale in Turkey. For more information, see https://kuzeyormanlari.org/hakkinda/ (accessed on 13 March 2022).

Fig. 2 Workshop on contract work for freelancers in IDEA Kadıköy in 2019, from the visual archive of Ofissizler

Fig. 3 Coworking day in Arthere, 2019, from the visual archive of *Ofissizler*. The space has immediate access to the street, and could host approximately ten coworkers

be missed a lot by the members of the network who had come together first to defeat isolation of working from home in the first place (Fig. 3).

Traditional demands on worker's rights rightly focus on job and social security, payment and working hours. As freelancers often lack even the most basic income security, the focus on that demand is obviously expedient. On the other hand, what is the ground from which to express those demands? Freelancers do not share the same space with colleagues, and hence often have little access to the knowledge and experience commons that could be created in the workplaces. Of course, there are serious impediments such as mobbing, hierarchical organization of work relations, and performance measuring systems promoting competition against such sharings in workplaces too. Gaining distance from such a disciplinary environment could have empowering impact on the worker. However, this is accompanied with a loss of access to means and relations of production, which they then are advised to procure by renting a spot in a coworking space. The primitive accumulation that happens here involves the enclosure of affective, social, and cognitive commons created and sustained by workers, their access to which is then limited through commodification.

So, thinking of coworking as a right to the city can be formulated among the demands of the precarious who may lack the material and immaterial resources that are needed to sustain the commons. Free access to such urban commons could contribute not only to the collectivization of means of production, but also class transformation as workers could seek and experiment with alternative ways of producing and living when they are thus empowered.

On the other hand, the existence of social infrastructure does not simply bring out such results. The way such spaces are framed and the kind of sociality they are based on is also crucial in assessing their potential to instigate such subjective and class transformations. To be more precise, one issue with public spaces could be their framing as a service provisioned by public authorities. The relation of service could frame the users of such spaces as recipients. The relation of the provider to the recipient here could also have a depoliticizing impact when and if the conditions in which the service is produced is not open to debate, and the recipients do not have a say in the management of the space. Such spaces could still be a ground of encounters, which could enable the flourishing of different relationalities. However, they could also have this depoliticizing impact in that it could provide anonymity rather than community to the recipient, limiting space to providing infrastructure for production.

That is why the flourishing of a post-fantasmatic relation to sociality in which class and non-class antagonism and social interdependence is rendered representable is crucial. Two important preconditions of such a representation would be the transparency relating to the conditions of production of such public spaces, and the participation of the precarious users of such spaces in the decision-making processes of their management and production. One final comment and line of research should be about the relation of such spaces with the larger context and access to urban commons and resources by the precarious in general. In a city like Istanbul, commuting and eating out could be very expensive for freelancers some of whom report not even earning as much as the minimum wage. The contradiction between the mobility they are promised and the impediments against its realization point to the need to approach urban commons from a class-based perspective as well.

Conclusion

The survey of three different sites of coworking in Istanbul provided here aimed to give an account of their political implications for precarious workers and their political organization. It is argued that private coworking spaces which are operated by capitalist firms employ a mutualist utilitarian discourse which involves a disavowal of class and non-class antagonisms and social interdependence. A form of social relation free from conflict is imagined and promised to potential users, who are advised to establish fruitful social relations by networking while also knowing that they can be disposed of any time they want. The sovereignty thus promised to freelancers among others promotes not autonomy but indifference with respect to one's spatio-temporal conditions and relations, which has serious ethico-political and affective implications.

On the other hand, it was suggested that commoning practices by freelancers, white-collar workers and the unemployed in *Dünyada Mekân* led to the emergence of several political tools of intervention into such capitalo-centric discourses and practices. It enabled the emergence and building of a network of solidarity among freelancers called *Ofissizler*, which aims to render problems of freelancing visible, formulate demands and imagine and enact alternative modes of production. However, it was also claimed that sustaining such spaces could be difficult due to the lack of mediating mechanisms which are needed both to manage the class and non-class antagonism and to establish links among different political tools so that they can form a quasi-uniform signifier that can hold participants together. Disintegration on the one hand, communitarianism on the other were mentioned as among the issues that could emerge otherwise. It was also suggested that precarious working conditions could make it difficult for freelancers among others to participate in such commoning practices.

Finally, the chapter proposed a thinking of coworking as among the right to the city, as part of the struggle around urban commons, and a public coworking space called IDEA in Kadıköy was critically mentioned in this context. However, it was also suggested that the framing of such spaces as service provision could have a depoliticizing impact as the consumer is turned into a recipient without any say in the conditions of the production of the space. Also, anonymity thus produced could impede interaction among users. Finally, the production relations productive of such spaces as well as the class processes playing out in them should be opened up to investigation to evaluate their transformative potential.

References

Avdikos, V., & Iliopoulou, E. (2019). Community-led coworking spaces: From co-location to collaboration and collectivization. In R. Gill, A. C. Pratt, & T. E. Virani (Eds.), *Creative hubs in question* (pp. 111–130). Palgrave Macmillan.

Bandinelli, C., & Gandini, A. (2019). Hubs vs networks in the creative economy: Towards a 'collaborative individualism'. In R. Gill, A. C. Pratt, & T.E. (Eds.), *Creative hubs in question* (pp. 89–110). Palgrave Macmillan.

Blagoev, B., Costas, J., & Kärreman, D. (2019). 'We are all herd animals': Community and organizationality in coworking spaces. *Organization, 26*(6), 894–916. https://doi.org/10.1177/1350508418821008

Brown, J. (2017). Curating the "third place"? Coworking and the mediation of creativity. *Geoforum, 82*, 112–126. https://doi.org/10.1016/j.geoforum.2017.04.006

Byrne, K., & Healy, S. (2006). Cooperative subjects: Towards a post-fantasmatic enjoyment of the economy. *Rethinking Marxism: A Journal of Economics, Culture & Society, 18*(2), 241–258. https://doi.org/10.1080/08935690600578919

Copjec, J. (2002). *Imagine there is no woman: Ethics and sublimation*. MIT Press.

Copjec, J. (2006). The object-gaze: Shame, Hejab, cinema. *Filozofski vestnik, 27*(2), 11–29.

David, I., & Toktamış, K. F. (Eds.). (2015). *Everywhere Taksim: Sowing the seeds for a new Turkey at Gezi* (pp. 185–196). Amsterdam University Press.

Derrida, J. (1995). *The gift of death*. University of Chicago Press.

Ehrensberger-Dow, M., Heeb, A. H., Massey, G., Meidert, U., Neumann, S., & Becker, H. (2016). An international survey of the ergonomics of professional translation. *ILCEA. Revue de l'Institut des langues et cultures d'Europe, Amérique, Afrique, Asie et Australie, 27*.

Enright, T., & Rossi, U. (2018). Ambivalence of the urban commons. In A. Ward et al. (Eds.), *Routledge handbook on spaces of urban politics* (pp. 35–46). Routledge.

Fink, B. (2003). Perversion. In M. A. Rothenberg, D. A. Foster, & S. Zizek (Eds.), *Perversion and the social relation* (pp. 38–67). Duke University Press.

Gandini, A. (2015). The rise of coworking spaces: A literature review. *Ephemera: Theory Politics in Organization, 15*(1), 193–205.

Gandini, A., & Cossu, A. (2021). The third wave of coworking: 'Neo-corporate' model versus 'resilient' practice. *European Journal of Cultural Studies, 24*(2), 430–447. https://doi.org/10.1177/13675494198860

Gibson-Graham, J. K. (1996). *The end of capitalism as we knew it*. University of Minnesota Press.

Gibson-Graham, J. K. (2006). *A post-capitalist politics*. University of Minnesota Press.

Gibson-Graham, J. K., Resnick, S., & Wolff, R. (2000). Class in a post-structuralist frame. In J. K. Gibson-Graham, S. Resnick, & R. Wolff (Eds.), *Class and its others* (pp. 1–22). University of Minnesota Press.

Gupta, A., & Ferguson, J. (1997). *Anthropological locations: Boundaries and grounds of a field science*. University of California Press.

Harvey, D. (2008). The right to the city. *New Left Review, 53*, 23–40.

İlyas, Ö. (2022). Freelance emek: Ofissiz çalışmanın sınıfsallığı [Freelance Labour: Class of officeless work]. Istanbul: İletişim Yayınları.

Ivaldi, S., Pais, I., & Scaratti, G. (2018). Coworking(s) in the plural: Coworking spaces and new ways of managing. In S. Taylor & S. Luckman (Eds.), *The new normal of working lives* (pp. 219–241). Springer.

Jakonen, M., Kivinen, N. H., Salovaara, P., & Hirkman, P. (2017). Towards an economy of encounters?: A critical study of affectual assemblages in coworking. *Scandinavian Journal of Management, 33*, 235–242. https://doi.org/10.1016/j.scaman.2017.10.003

Kojo, I., & Nenonen, S. (2016). Typologies for co-working spaces in Finland–What and how? *Facilities, 34*(5/6), 302–313. https://doi.org/10.1108/F-08-2014-0066

Madra, Y. M., & Özselçuk, C. (2010). Jouissance and antagonism in the forms of the commune: A critique of biopolitical subjectivity. *Rethinking Marxism: A Journal of Economics, Culture & Society, 22*(3), 481–497. https://doi.org/10.1080/08935696.2010.490409
Madra, Y. M., & Özselçuk, C. (2015). Creating spaces for communism: Postcapitalist desire in Hong Kong, The Philippines, and Western Massachusetts. In G. Roelvink, K. St. Martin, & J. K. Gibson-Graham (Eds.), *Making other worlds possible: Performing diverse economies* (pp. 127–152). University of Minnesota Press.
Merkel, J. (2019). 'Freelance isn't free.' Co-working as a critical urban practice to cope with informality in creative labour markets. *Urban Studies, 56*(3), 526–547. https://doi.org/10.1177/0042098018782374
Osteen, M. (2010). Jazzing the gift: Improvisation, reciprocity, excess. *Rethinking Marxism: A Journal of Economics, Culture & Society, 22*(4), 569–580. https://doi.org/10.1080/08935696.2010.510303
Papageorgiou, A. (2016). *Living in 'Beta': Hubs, collectives and the emergence of moral economies in Athens* (Dynamics of virtual work, Working paper series 7). Available at: http://dynamicsofvirtualwork.com/wp-content/uploads/2015/08/working-paper-No-7.pdf
Resnick, S., & Wolff, R. (1987). *Knowledge and class: A marxian critique of political economy*. University of Chicago Press.
Spinuzzi, C., Bodrožić, Z., Scaratti, G., & Ivaldi, S. (2019). "Coworking is about community": But what is "community" in coworking? *Journal of Business and Technical Communication, 33*(2), 112–140. https://doi.org/10.1177/1050651918816357
Tintiangko, J., & Soriano, C. R. (2020). Coworking spaces in the global south: Local articulations and imaginaries. *Journal of Urban Technology, 27*(1), 67–85. https://doi.org/10.1080/10630732.2019.1696144

Self-Organised Labour: Reclaiming the Commons in the Post-Socialist Space of Southeast Europe

Dimitra Gkitsa

Introduction

The current neoliberal state of social inequality, precarity and generalised crisis has brought to the front a revival of the commons. Initially, the commons as a term referred to the uncultivated fields and lands around town and villages that allowed the local communities the right to access natural resources. Responding to Hardin's 'tragedy of the commons' (1968) and his argument that the only way to avoid the destruction of natural commons was through government regulations or privatisation, Elinor Ostrom (1990) reintroduced the concept of the commons to highlight that there were many cases where communities succeeded in sustainably managing the natural commons without state regulation or private property regimes. Commoners, the collective subject that governed the commons, were self-reliant, had access to diverse sets of natural resources and this independence was paramount to recognise and to establish their political power. The commons here provided the very basic elements and resources that any human being needs to ensure its survival, and at the same time it implied the first political arrangements around constructing and constituting communities. In this aspect, ensuring the continuity of the commons meant ensuring the survival of the community. Even during the first peasant communities, the commons did not simply encompass a form of distribution or resources or even a production system, but combined a nexus of social behaviours, everyday norms, cooperation, and collective decision-making (Federici, 2004). The commons were also spaces of public gatherings, celebrations, rituals, ceremonies, and pagan festivities (Perelman, 2001; Linebaugh, 2008). In the mid-sixteenth century, a series of privatisations eliminated communal land, threatening the survival

D. Gkitsa (✉)
School of Slavonic and East European Studies, University College London (UCL), London, UK
e-mail: d.gkitsa@ucl.ac.uk

J. Merkel et al. (eds.), *Coworking Spaces*, https://doi.org/10.1007/978-3-031-42268-3_3

and the mere existence of the commons as they had been known until then. The enclosures, a term defining the expropriation of common and communal land, brought dislocation of communities and introduced the very first conditions of waged labour, which was a decisive element for the later development of capitalism (Hardt & Negri, 2009).

The rapid transformations that occurred during the past years in neoliberal cities and towns, described as a 'dispossession by accumulation' (Harvey, 2007),[1] have led to a revival of both theoretical conversations on the commons as well as its practical applications within urban spaces. The concept of accumulation by dispossession defines the global neoliberal changes that were established in the Western world from the 1970s to the present day. Rapid privatisations, intense financialisation, management and manipulation of crises, and nation-state redistribution are the very practices through which neoliberalism takes concrete forms. Amid this reality, the commons have re-emerged as a social movement, as an urban emancipation, and as a horizon (De Angelis, 2017; Stavrides, 2019). The latter refers to re-imagining and making possible (infra)structures that contain more equal and fairer forms of working and being in the present neoliberal reality while also challenging this reality. In this aspect, self-organisation is at the core of commoning (the very act of practising the commons) as it enables collective control and access of resources, distributed power, collaborative decision-making, and in some cases, reclaiming of actual spaces that have been taken away from their original communities.

The current state of precarity, digital economy as well as the increase of project-based labour performed by freelance and self-employed workers has led to the appearance of alternative shared working environments which offer the physical space as well as the material and immaterial resources that are vital to the labour of independent workers. Although coworking spaces have been understood as social hubs that are able to facilitate collaborative activity (Spinuzzi, 2012), they have not managed to escape commercialisation (Avdikos & Merkel, 2019). The fact that private companies step in to profit from the lack of available communal spaces and resources highlights both the solution and the symptom of the problem. Many authors have pointed out the shifting nature of labour in the dawn of late capitalism (Lazzarato, 1996; Virno, 1996; Hardt & Negri, 2001; Terranova, 2004; Caffentzis & Federici, 2014). This stage of capitalism is based predominantly on immaterial and affective labour whilst the very neoliberal power that sustains such labour has turned out to be more fluid and decentralised. As such, it has also become more difficult to

[1] The concept of accumulation by dispossession describes for David Harvey the neoliberal practices that were established in many parts of the Western world from the 1970s to the present day. Harvey identifies four crucial practices that come to define the process through which neoliberalism takes concrete form: intense and rapid privatisations, financialisaton, management and manipulation of crises, and nation-state redistribution. These four practices have become central to the functioning of capitalism under neoliberal policies (Harvey, 2007). Particularly, the accumulation of power and centralisation of wealth finds its resonance in the structure of urban space.

address the state and conditions of precarity collectively.[2] How can we generate new spaces for collaboration and cooperation in this reality? More crucially, what are the emancipatory potentials of self-organised labour?

Waged labour has been crucial to capitalist production. Within Marxist tradition, the worker's labour power is understood as a commodity that is exchanged for wage. While labour involves the activity of work, labour power involves the human abilities (both physical and intellectual) that are essential in performing work. However, post-Fordist production, a term that describes the conditions of contemporary work in global capitalism, has blurred the previously distinct connection between labour power and wages. Post-Fordist production is characterised by the dominance of the global market, the shift from manual labour to the service sector, the disrupted linear relation between employer and employee, and an ever-increasing precarity. At the same time, the gradual loss of previously established welfare rights and benefits, exploitation, unemployment, the normalisation of temporary employment or unwaged labour justified through public rhetoric of crisis and austerity are elements that persist in present capital-labour relations. This reality does not stop within the temporal and spatial limits of our workspaces and work conditions, but it also expands in other areas of public and private life, shaping our subjectivity, our everyday life, our social interactions and even the ways in which we understand and spend our free or leisure time (Frayne, 2015). As such, many thinkers understand labour to be directly linked with bio-politics (Berardi, 2015; Hardt & Negri, 2009; Mitropoulos, 2012). A conversation then on the commons inevitably invites us to reconsider alternative forms of organising labour. In fact, the commons rely on self-organised labour that is in its essence a collaborative practice. The individual labour is transformed into a collective labour that can challenge and, in some cases, transcend conventional capital relations, giving control of labour power to the collective body of the workers who can define themselves the purpose and the conditions of their labour.

This chapter explores collaborative practices and coworking spaces using as an entry point self-organised initiatives in the cultural sector in Southeast Europe to understand the emancipatory dimensions of collaborative work. In analysing the ways in which self-organisation produces infrastructures and networks of solidarity by commoning the post-socialist space, this chapter argues that collective labour contains an element of affective political action. In this sense, workspaces can be approached as spaces for practising the commons. Whilst self-organisation has emerged within the western world as an alternative to hegemonic practices and precarious labour, the former Yugoslavia reached that ideal, but its legacies have been marked by a contested collective memory of the socialist past. Here, we do not simply speak about a horizon, but rather, about the memory of a horizon that was once reached, then abused, and finally lost.

[2] For instance, forms of remote labour that were established during the Covid-19 pandemic, rendered traditional work offices and physical spaces redundant, showing that labour could be carried on as usual from anywhere even amid a global health crisis.

Remembering the Commons: A Brief Overview of the Workers' Self-Management in the Former Yugoslavia

The main difference between the history of the commons in Southeast Europe and the West is the experience of the worker's self-management in the former Yugoslavia. Breaking up connections with the Soviet Union in the early 1950s, Yugoslavia became politically isolated from the state-socialist East as well as the market-capitalist West. A new socio-economic governing system had to be invented, one that would be based both on anti-capitalist and anti-statist ideology. Workers' self-management[3] introduced the gradual withdrawal of the state and advocated for its replacement with a mechanism in which the workers would be in control of their labour through councils. Through professional councils, workers could participate in the management and distribution of resources and to negotiate their working conditions. In the arts and cultural sectors, councils allowed creative practitioners opportunities such as pension schemes, participation in exhibitions and studio spaces.

By the 1970s the workers' self-management system started having organisational and functional problems. This was mainly because of the complicated bureaucracy which allowed the upper-class technocrats who possessed a better knowledge of resources and had developed links with banks and local authorities, to accumulate power. Nevertheless, workers managed to question the very foundations of the self-management system within the professional associations that were born out of it. Indeed, as the Slovenian philosopher Gal Kirn (2019) points out, although self-management was initiated from top-down and had distinct ideological purposes, it can also be read as a movement from below:

> which speaks of successful and long-term practices of organising workers in specific companies and social/cultural sectors, as well as in a more antagonistic sense of the state, in the form of illegal self-organisations in strikes and protests that challenged the Party hegemony (p. 98).

Workers did not initiate the self-management system themselves. However, this social experiment created room for vigorous ideological negotiations of the very conditions of work. Commenting on Ljubljana's alternative cultural and student movements of the 1980s, Marina Gržinić (2008) explains that it 'went beyond the counter-cultural attitude of the 1970s, demanding new cultural, political and artistic institutions and organisations to be formed, so to speak, within the very institutions of the socialist self-management paradigm of reality' (p. 97). It was amidst this political reality that artistic and cultural collaborative practices started to emerge.[4]

[3] The system of workers' self-management was initiated and developed by Edvard Kardelj, the vice-president of Tito's government and former partisan during the Second World War.

[4] For instance, artist groups such as IRWIN who co-founded the larger collective known as Neue Slowenische Kunst; the Belgrade Six in Serbia, a generation of conceptual artists who graduated from the Academy of Fine Arts in Belgrade, the Grupa Sestorice in Zagreb; and the A3 group (1970–1973), who initiated public intervention in urban spaces in Belgrade.

Similarly, Students' Centres in Zagreb, in Novi Sad, in Ljubljana and in Belgrade were hubs for artistic and cultural freedom amongst young artists and creative workers. The art practices produced within those frameworks were not just a representation of a new approach to art making, but rather 'a new form of organisation with state-supported art institutions' (Jakovljević, 2016, p. 28). This was most radically brought forward in *Oktobar 75*,[5] during which artists directly addressed the status of their labour in the sphere of the political and social conditions within Yugoslavia. As such, it was within these collective groups that artists managed to build resistance and critique against the state. In addition to pushing the boundaries of artistic experimentation and critique towards the Party, the self-management system offered the socio-political environment for cultural workers to contemplate their work production as labour, to experiment with the very ways they organise and access resources, and as such, to survive in a reality where freedom of expression was strongly controlled. Organising independent groups beyond the self-organised workers enterprises and adopting an activist character in art practices, reflect the constant struggles for autonomy in the cultural field.

Self-management in former Yugoslavia remains, even nowadays, poorly researched, as it is strongly associated with its socialist legacy.[6] The last decade of the twentieth century in Southeast Europe was marked by the dismantling of the socialist heritage, intense privatisation, and implementation of market reforms. The years of transition from socialism to neoliberal democracy were experienced as an abrupt cultural, social, and political shift in the ways with which public space and property were understood. Transition became a nuanced ideological term which justified a series of structural and institutional reformations intended 'to produce not only a more prosperous economy, but also a whole new political and economic management structure that would lead these societies towards democracy' (Pusca, 2007, p. 32). A sudden transition from one status quo to another, manifests itself in all aspects of the collective and individual realms of the everyday, the mundane and the ordinary, demanding a radical shift in the ways people live, work, interact with each other, and understand their existence in the world.

The collapse of the socialist regime in the Yugoslav space was also followed in the 1990s by brutal wars and massacres. Ethnic tensions between countries which once were united under the same socialist identity, brought to the surface new issues

[5] October Salon was founded back in the 1960s by the City of Belgrade remaining since then a reference point in the visual arts and its new tendencies in Serbia.

[6] It is interesting to note here that even terms such as "cooperative" (*zadruga*) have nowadays negative connotations because they are associated with the socialist past (Tomašević, et al., 2018). Initially appearing as a term in the early 19th century, *zadruga* referred to a rural form of familial agricultural cooperative that was spread in the Balkans (mainly in the Slavic countries and Albania) from the Middle Ages and existed until the early 20th century. The historian Maria Todorova understood *zadruga* both as a structure and as a process "possessing a number of diverse valences, such as kinship, property relations, residence, working arrangement" (1990, p. 42). Theterm was later re-introduced by the Communist Party of Yugoslavia to describe its collectivisation of agriculture through which private lands and individual agricultural labour were turned into collective farms and cooperatives.

of nationalism, xenophobia, and the revival of ancient quarrels over borders. Remembering former revolutionary moments within Yugoslavia such as the workers' self-management, and revisiting collectivism in the post-Yugoslav space, is for many cultural practitioners a strategy to build connections of solidarity that go beyond the rhetoric of nationalism(s) that followed the wars. At the same time, self-organisation and DIY practices in the cultural sector appeared as the only possible tactic to keep working in a reality where institutions were practically dismantled and public spaces as well as resources had been transformed into private corporations. In the West, collective autonomy became a model for cultural workers to survive outside of or in parallel to the market and established institutions. Until recently, cultural workers in the post-socialist countries had neither of these. Instead, what they had was an inherited collective ethos and a prior common knowledge for forming collaborative work models in the present. In the social, political, and cultural reality of the deconstructed post-socialist space, the struggle around the commons is strongly anchored with the socialist past, its social structures in the everyday as well as the material remnants in liminal public spaces.

Practising the Commons After Socialism

The commons, or the process of doing the commons, has been articulated by scholars and activists as a social praxis and action. This includes the continuous process to reclaim spaces, build supporting communities and share both immaterial and material resources. This approach suggests that the commons do not exist per se, but rather they are practised, pursued, and produced collectively. The activist and writer David Bollier (2008) mentions that the commons revolve 'around collective participation and shared values. [...] Generically speaking, a commons is a governance regime for managing collective resources sustainable and equitably' (p. 144). These resources, in addition to the material world and actual spaces include also the immaterial 'results of social production that are necessary for social interaction and further production, such as knowledges, languages, codes, information, affects and so forth' (Hardt & Negri, 2009, p. viii). In addition to common shared resources (or better, a common struggle to reclaim resources) and the self-organisation of social relationships and means of production, the commons involve a third factor: a community of participating members, the very subject of the commons. As De Angelis highlights, we need to recognise the commons as a verb—to common (An Architektur, 2010). This means that the commons is a constant practice, a continuous action.

Collaborative labour is at the very foundations of the commons. What forms can this collaboration take and in what ways can collaborative labour shift working spaces and ethics? The case studies presented below show that collaborative work practices are in their essence heterogenous, developed organically in accordance with the needs, the demands, and the local struggles of their retrospective communities. They are also distinct when it comes to operational and structural elements. For instance, Rojc Community Centre is a self-organised network of individual organisations, Magacin is a self-organised coworking space and Termokiss is a social space run by its community. And yet, they equally demonstrate distinct similarities when it comes to practising the commons. Firstly, they all adopt a horizontal work model that is based on continuous communication and participation in the decision-making process. Secondly, they also carry a similar collective effort to reclaim communal spaces against privatisations and neoliberal institutional policies that have taken away social spaces and public resources.

Rojc Community Centre

Rojc Community Centre located in Pula, Croatia consists of over one hundred individual organisations that all operate within the premises of Rojc, the largest building in the town which covers a surface of 16,739 square metres. Rojc's current organisations cover a range of activities in the cultural and social sectors. The building was constructed in 1870, under the Austro-Hungarian monarchy and it was initially used as a naval military school. After the Second World War the building was renamed after the partisan hero Karlo Rojc and it was used initially as an engineering school and later in 1976 it became a military base. In 1991, when the wars erupted, the building accommodated refugees. In the late 1990s the building was completely abandoned and decayed. It was during this time when individual cultural organisations, because of lack of available spaces, reinhabited the building, restored some of the common areas and started organising community activities.

Rojc is a unique example of how a collective initiative, which started basically as squatting, turned into a formally recognised entity. This long process started in 1997 when the municipality, unable to avoid the fact that the Rojc's premises were being used, offered use of the building to the organisations in exchange for rent. The City of Pula had initially agreed to pay for the maintenance of the building. However, this caused more conflicts as the process was not made transparent and was followed by mismanagement of funds. In 2008, the organisations reacted with a large public protest, demanding to be part of the decision-making process about issues related to the building. The result of this was the 'Coordination of the Community Centre Rojc', which led to co-management agreement between the Rojc organisations and the City of Pula. However, the unwillingness of the municipal authorities to actively include Rojc members in decisions related to the management of the building resulted in more conflicts. In 2009, negotiations that culminated in a protest, led to an agreement of full partnership rights with the City. Decisions related to the operation of Rojc were taken by mutual agreements between three Rojc representatives and three representatives of the city, who are appointed by the mayor of Pula. Rojc's organisations through their self-organised bodies monitor expenditures, to coordinate issues related to maintaining the building and to collectively negotiate with the City of Pula. On the other hand, the City of Pula has committed to an allocation of annual funds for the maintenance of Rojc. They also cover salaries for night guards, cleaning services, insurance, and accounting.

All users of Rojc can actively participate in establishing the rules for the usage of the space, but the City of Pula remains the official owner of the building and a city-owned company oversees its maintenance. As the organisations are required to pay rents, to constantly negotiate their contracts with the City of Pula, and to seek agreements with the local governmental body, tensions and struggles are ongoing. However, by forming a network the organisations managed to build a civic-public partnership with the City of Pula, co-governing Rojc. The notion of space in this case is particularly important considering that privatisations and complicated bureaucratic procedures with local authorities have made access to spaces even more difficult for cultural organisations. Space, and having access to space, is a vital resource for the actual work conducted by the cultural organisations. But more than this, space in the case of Rojc is also returned to the broader community of Pula. This takes place not only with the very character of labour at Rojc which includes activities and services that benefit their society, but also with keeping the space open and accessible to its communities to use. For instance, the 'Living Room' is a designated space in Rojc that is open both for the organisations in Pula to accommodate their public events and for any member of the public (Tomašević et al., 2018, p. 87). To this end, I would argue that more than a public space, Rojc has turned into a common space. The commons in the case of Rojc are practised not only on an internal level with regulations that allow a direct participation to its users, but also on an external to the organisation level by the participation of a broader public community which can use spaces in Rojc. As the Care Collective (2020) mentioned in their recent publication, 'public spaces are crucial for building caring communities because they are egalitarian and accessible to all, and can foster conviviality, interconnections, and the emergence of communal life. We must create, take back and demand more public space' (p. 49). The ongoing tensions and negotiations with the City of Pula is a process of reclaiming and making space.

Although the City of Pula remains the final power as it approves the propositions put forward by Rojc, organisations that are part of Pula have gained a significant representation and visibility by establishing a cross-organisational grassroot alliance and solidarity. Self-organisation in this case covers the whole spectrum of operations within the building of Rojc and not just that of a single organisation. Exactly because Rojc involves a complex ecosystem of individual organisations, disagreements amongst its members are inevitably part of such a self-organised model (Mišković, 2018, pp. 185–186). Constant conversations, negotiations, and communication, both internally and externally, are particularly important. Similarly important is also the horizontal governing model because it offers the space for individual opinions and interests to be heard. This highlights that a community is never harmonic or homogenous. On the contrary, a community is constantly done and undone by its members, remaining in a sense open to change and to develop organically.

Following Chantal Mouffe's (2013) observations that every hegemonic order is 'susceptible to being challenged by counter-hegemonic practices' (p. 2), I would argue that Rojc's practice of self-organisation is an agonistic counter-hegemonic action. For Mouffe, cultural and artistic practices contain the ability to 'foster agonistic public spaces where counter-hegemonic struggles could be launched against neoliberal hegemony' (2013, p. xvii). Rojc's self-organised practice is not only counter-hegemonic in the sense that they are in constant struggle with existing formal power structures (in this case the City of Pula), but more crucially, is it also counter-hegemonic in the sense that they operate with alternative modalities of labour and cooperation. Contrary to hierarchical cultural institutions, users of Rojc have a direct participation in decisions related both to their retrospective organisation and to the larger Rojc community.

Magacin Cultural Centre

While Rojc's self-organisation receives the form of a broader alliance and a network of solidarity between organisations, Magacin is a cultural and social centre that works essentially as an open platform and a coworking space for creative practitioners and cultural workers. It was formed in 2007 in an abandoned former publishing warehouse in Belgrade. Magacin functions with the principle of an 'open calendar model', which means that facilities, recourses, equipment, and spaces in the buildings are openly available and accessible to anyone (organisations, collectives, or individuals from the fields of art, culture, and social activism) free of charge. Essentially, this means that Magacin is a platform that gathers the material and immaterial resources (including actual space as well as equipment) that the members of the community need for their creative work. The main principle is that those who use the space cannot be organising events for commercial purposes, such as advertising sponsors or charging fees for the activities they offer. Magacin consists of three different buildings located on Kraljevića Marka Street. The first building contains a large central space (which can be used for various activities such as rehearsals, performances, film screenings and exhibitions), a dance hall, a coworking space for joint office work, a meeting room, a cinema, a basement space which is used for reading and theatrical rehearsals and a room for storing technical equipment. The second building is an open workshop which contains tools and workstations for creative labour that incorporates metal, wood, electronics, textiles, and printing. It is a space which artists and creative practitioners can use. A third building is used predominantly for exhibitions.

Magacin is self-defined as 'a resource managed by the community of users according to the commonly established rules' (Čukić et al., 2020, p. 120). According to the centre's statement, these collectively established agreements, which more than rules are common values/ethics related to their labour include: (1) responsibility in managing and establishing the public programmes and activities; (2) cooperation and relationship of mutual trust amongst the members; (3) a commitment for making all activities accessible and free of charge to participants; (4) financial transparency; (5) respect of personal and artistic autonomy; (6) a commitment to social change for the benefit of a larger group of people (Magacin, n.d.). Through their horizontal and participatory governing model, Magacin has managed to build an ecosystem of solidarity. Like Rojc, Magacin also reclaimed an abandoned building and has offered it back to its community. The interesting difference here is that while Rojc is a network of social and cultural organisations, Magacin is an open space that can be practically inhabited by any creative worker, collective or individual organisation. The need to create this space was the present reality in the cultural sector in Serbia and a lack of adequate infrastructure that would support the conditions of creative labour. These include both material and physical elements of labour (space, funds, resources), as it equally implies the cognitive capacities and the knowledge production that occurs in the different forms of social cooperation (discussions, collaborations, collective narrations). Cultural workers have turned into this new type of working class, which Kuba Szreder calls 'the projectariat'. Szreder (2021) defines the projectariat as:

> people who, in order to survive, have no choice but to make and implement projects, as many of them do not own much beyond their capacity to move in artistic circulation. In this situation, they are both similar and distinct from proletarians, who have to sell their own labour force in order to survive. Similar to proletarians, projectarians are forced to chase temporary possibilities of employment provided by projects or jobs structured as if they were projects (i.e., temporary and task-oriented assignments). On the other hand, projectarians are entrepreneurs of the self, because projects enable them to capitalise their skills, social connections and reputations in exchange for opportunities and other future gains (p. 194).

This highlights on the one hand the normalised state of precarity, and on the other hand the shifting nature of immaterial labour. This type of labour which is heavily dependent on the production, reproduction, and maintenance of social relations and networks, is at the very core of the new spirit of capitalism (Boltanski & Chiapello, 2007). Contemporary capitalism, exactly because it is based on immaterial forms of social exchange, produces more subtle forms of exploitation which in turn perpetuate existing unfair work cultures.

In reference to the above reality, Magacin appeared in an attempt to provide to creative practitioners and cultural workers a space where they can produce and showcase their work. This collective effort does not come without struggles. Magacin's legal status remains even nowadays unresolved. Although the City of Belgrade had at first endorsed the initiative, the new city coalition after the 2008 elections did not support the idea of having a social centre in the city. This escalated when the users of Magacin were ordered to leave the building in 2014. They never left. What they did is to put forward a new governance model for the usage of the space. The managing bodies of Magacin are the Users' Assembly and the Coordinating Body. The User's Assembly manages the space, making decisions related to Magacin's buildings, the activities held in them, its maintenance, and conditions of use for its community. On the other hand, the Coordinating Body is responsible for advocating as well as for securing and managing funds.

The commons in Magacin are practised via regular meetings amongst the users of Magacin which offer the space and the circumstances for decisions to be taken collectively and in a horizontal democratic way. At the same time, essentially any member of the local community can become a long- or short-term user if they agree to the conditions of space usage. The users, which in this case are both cultural organisations that might need a space and individual workers, also consist of the community of Magacin. Magacin has turned into a hub where creative and cultural workers have a space to work together and to share resources that are necessary for the production of their labour. Workers who use the premises and the resources of Magacin, more than having a working space, also automatically become part of a larger community. This allows them to actively participate in dialogues and decisions related to the improvement of their work conditions in the space of Magacin and to cooperate with fellow workers whilst maintaining autonomy of their own labour and creative production. Precarious cultural workers and creative practitioners have a platform to keep existing outside the dominant hierarchical structures of institutions and without market interventions, demonstrating that there can be another way of working and existing in the contemporary art and cultural sector. More crucially, labour is understood as a vehicle that could nurture social change.

Termokiss

Termokiss is a community centre in Pristina which started its operations in 2016 when the NGO organisation Toestand[7] from Belgium in collaboration with Info Quartier Mache from Switzerland and Kosovo Architecture Festival as well as numerous volunteers, local citizens and cultural activists in Kosovo reclaimed and revitalised a building in Pristina. The building, a vacant industrial space, was meant to be completed in the late 1980s to be used by the heating company Termokos. However, due to conflicts between Kosovo and Yugoslavia, the project remained unfinished. The ownership of the building came to the hands of the Municipality of Pristina which left the building in a state of complete abandonment. The main difference between Termokiss and Rojc or Magacin is that Termokiss involved in its creation foreign organisational entities which offered the practical knowledge as well as the basic first resources to the local community to set up a self-organised social centre. Thus, they responded in a way to the pre-existing need for having a social space that would truly belong to its community. The public space of Kosovo in addition to having undergone privatisations and regeneration projects that have taken away the public space from its communities, still bears the marks of the war. In fact, the urban space of Pristina, demonstrates with its infrastructure all the transitions and the struggles that comes with the creation of a new country.

[7] Toestand is an NGO that as part of their activities collaborate with local organisations and volunteers across Europe to reclaim and to reuse abandoned buildings in their neighbourhoods. Like Termokiss in Pristina, the NGO has also helped setting up the self-organised social centres SCS Tetova in North Macedonia and Uzina in Tirana, Albania.

Termokiss was the outcome of a six-month research project during which interviews were conducted with citizens of Pristina who were asked about their needs and the ways in which potential infrastructure in the city could cover those needs. The building which covers a two-hundred square metres usable space includes a co-working space, library, kitchen, storage room, bar as well as an outdoor garden with sports facilities and a playground for kids. It has become a centre of public gathering. Termokiss organises regular cultural activities, including workshops, concerts, and screenings. However, the space remains open for anyone who would want to use it under the condition that it will not be used for commercial purposes, but rather for the benefit of the wider community. This becomes even more evident considering how the space became usable through the work and care of volunteers. Members of the community were involved in the first steps of forming the centre and deciding its missions through participating in regular open meetings.

Since Termokiss was a joint effort between a foreign NGO and numerous local volunteers, I would argue that Termokiss is a form of self-organisation that appeared as the outcome of a 'mutual aid project'. Distinguishing mutual aid projects from charity work, lawyer, and activist Dean Spade (2020) writes that:

> Most mutual aid projects are volunteer-based and avoid careerism, business approach, and charity model of nonprofits. Mutual aid projects strive to include lots of people, rather than just a few people who have been declared 'experts' or 'professionals'. If we want to provide survival support to as many people as possible for root-causes change, we need to let people do the work and make decisions about work together, rather than bottlenecking the process with hierarchies that let only a few people lead (p. 59).

What appears from the above statement is an important predicament about the role that non-profit organisations play in communities that need help in forming sustainable infrastructure. In this process, once again the horizontal involvement of the local communities when it comes to decisions and organisation is fundamental. Termokiss received this type of aid in its initial formations which was paramount in the process of negotiating with the Municipality of Pristina and in securing funding from volunteers and individual donors. For the Municipality to recognise Termokiss as a social centre they had to register as a new non-profit organisation. Since this type of organisation still implies a form of governance based on hierarchies, the non-profit element appears only on the papers. The actual operation and management of the space is conducted by its community in a horizontal participation. The non-profit legal status has helped Termokiss to secure agreement with the Municipality which has handed complete rights over the building and the green area surrounding it. In its current form Termokiss is governed by what they call Termokiss Platform. This includes the Termokiss Community, which is responsible for general decision-making, and the Termokiss Space, which oversees the events that happen in dialogue with the needs of the community. The community in this case are mainly younger people who have stepped in to create themselves a social space where they can come together to think, create, work, and spend leisure time. Termokiss also includes different working groups, called 'the clubs', which are developed and shaped according to the needs, ideas, and interests of members of the community.

The main challenge of Termokiss is to maintain its financial stability since operations are taking place on a volunteer basis. In this case, we speak about voluntary labour that fills up what would have been someone's free time. Termokiss is a self-organised initiative that practises the commons through horizontal governing structure and through the creation of a space that is open and available for everyone. Volunteer labour is what sustains and reproduces the commons. Contrary to unpaid labour, in the case of Termokiss, volunteer energy and time become the vehicle through which citizens can participate in public life, engaging with issues related to their community, and have a direct participation in collective activities which are essential parts of any society.

An Affective Political Power

The philosopher Boris Buden (2007) reminds us that institutions in East Europe after the collapse of socialist regimes adopted a neoliberal model that was shaped and constituted by the classical ideology of liberalism, following the economic and political paradigm of global capitalism. This stage of late capitalism is based predominantly on affective and immaterial labour. Affective labour produces immaterial goods such as services, emotions, knowledge, information, and social networks. Similarly, the decentralisation of work time and workspace means that the distinction between life and labour have also become blurry. It occurs then that 'capitalism is not just an abstract inhuman agency "out there", instantiated in forms of technology, and so forth (that is, as a supra-molar entity). It is also "in here"—producing our very subjectivity on what we might call a molecular level' (O'Sullivan, 2014). This means that capitalism operates not only at a macro-political level, but also on a bio-political level, shaping our subjectivity, our bodies, the rituals of our everyday lives, the ways we remember and collaborate, and our social relationships. If the commons have reappeared in the neoliberal city as counter-hegemonic movement against a reality where capital is produced and reproduced by and through affective labour, then I would argue that a conversation on the commons—on the politics of practising the commons—should equally be a conversation on the politics of affect.

More than just an emotion or a feeling, affect has been articulated as the ability that everyone and everything carries to affect and be affected (Gregg & Seigworth, 2010, p. 2). In this sense, affect is the process of becoming (Deleuze & Guattari, 2004, c1987) and as such affective encounters contain powerful transformative potentialities. To this end, collaborative spaces, practices and inevitably, politics are not just about working together, but rather they contain the affective potentiality if not to entirely shift work cultures and given realities, then at least to create the imaginary of a different alternative. The case studies of self-organised initiatives analysed in this chapter are essentially based on affective labour. This self-organised labour is affective not only because its products are social spaces and platforms for social gatherings, but also because it is based on practices of care; care for both the actual workers who are involved in maintaining their self-organised structures and for their broader communities. To this end, I would argue that commoning practices contain an affective political power. They are affective because although they might not be sufficient to change practices on a broader institutional level, they are still able to intervene in the micro-politics of the everyday life and its social formations and interactions. The affective political power that comes with self-organised labour contains an emancipatory rupture in the sense that it does not only simply disturb hegemonic power relations in their local settings, but also its shows that there is still space and time for generating alternative modalities of (co)working and (co)existing in the present reality.

The affect theory scholar Lauren Berlant (2016) notes that 'the commons is an action concept that acknowledges a broken world and the survival ethics of a transformational infrastructure. This involves using the spaces of alterity within ambivalence' (p. 399). Berlant (2016) detects in affective infrastructures a kind of movement and flexibility that does not exist in institutional frameworks as they still enclose and gather around specific discourses and forms of power relations. While infrastructures function in the domain of habits, everyday norms and informal patterns, institutions 'represent their legitimacy in the way they represent something reliable in the social, a predictability on which the social relies' (Berlant, 2016, p. 403). When institutions represent authority and official truths, affective infrastructures offer glitches that could be commoned and transformed into something different, including an alternative way of organising and understanding the very essence of labour.

In the specific case of the former Yugoslav space, which was the main point of reference for this chapter, the promised neoliberal future instead of the much-desired prosperity has brought additional precarity and social struggles. Here, the past keeps providing those glitches, taking the form of both the forgotten and abandoned actual spaces that are inhabited once again and the previous common ethos of collectivism which has been transmitted to the younger generation of activists and cultural practitioners. As art curator Ana Dević (2021) writes in a recent publication on collectivism in the post-Yugoslav region, 'the generation of artists born in the early 1990s, when the former Yugoslavia was riven by genocidal nationalist wars, will probably be the last generation to be touched by the legacy of socialism—not through personal memory, but through remnants and traces of socialist architecture, history, and political values'. I would add to the above that this includes not only artists but a whole group of precarious workers in the civic society and in the cultural sector who are turning once again into collaborative practices to ensure their collective survival.

Conclusion

The post-socialist condition, arriving with the failure of the communist projects in Eastern Europe, announced an era that is characterised by its inability to produce social and unifying imaginaries, haunted by the loss of shared time and space. Mark

Fisher (2014) writes that 'anachronism and inertia' are the main characteristics that describe the experiences of 21st century culture (p. 6). Whilst past social movements were able to negotiate labour conditions and to advocate for fairer wages, the current reality of precarity, unemployment, unstable time-based or project-based jobs, and the absence of concrete workspaces[8] has made it even more difficult to translate the generalised struggle into a common and unifying movement. In the cultural sector, workers are expected to be multi-skilled flexible entrepreneurs of their own services, or in other words, to operate as self-monitored 'microstructures' (McRobbie, 2002). Interestingly, such traits of precarious work are similar both to freelancers and to workers employed in organisations and institutions (Child et al., 2017, p. 153). Whilst workers are expected to be able to adjust to the rapidly shifting conditions of our present reality, institutional structures are hesitant and slow to incorporate change. However, even amid this reality 'the long, dark night of the end of history has to be grasped as an enormous opportunity' (Fisher, 2009, p. 80). The question that arises then, is what kind of shapes that opportunity can take? This chapter has shown that this opportunity needs to be a collaborative project.

Collaborative practice, more than just a condition of working together (inevitably, we 'work together' in any type of labour that we perform), is an emancipatory project that contains a collective consensus to self-organise. To this end, collaborative practices produce affective infrastructures, including actual spaces, resources, alternative knowledge production as well as social relations. Infrastructures that emerge through common practices can generate an additional liminal space that stands in-between the neoliberal markets as well as the bureaucracy and hierarchy of existing institutions. Collaborative practices do not come without struggles. They require not only painful negotiations with external authorities, but also constant fostering of relationships amongst the diverse individuals that make up the community of self-organised workers. Such emotional efforts in many cases remain invisible and as such, are translated into unpaid labour. However, the many self-organised initiatives that exist in the world demonstrate that collaborative working spaces and practices can be formed, sustained, produced, and reproduced by their very own community of users and without the invisible hand of the market or the bureaucratic regulatory interventions of governmental authorities.

[8] The most characteristic examples of this would be workers employed in the gig economy as well as workers of companies which operate as online platforms. Such business models that dominate the global market have disrupted and blurred traditional relationships between employers and employees, operating with exploitative practices and unregulated control (see Srnicek, 2017; De Ruyter and Brown, 2019).

References

An Architektur. (2010). On the commons: A public interview with Massimo de Angelis and Stavros Stavrides. *E-flux, 17*. https://www.e-flux.com/journal/17/67351/on-the-commons-a-public-interview-with-massimo-de-angelis-and-stavros-stavrides/

Avdikos, V., & Merkel, J. (2019). Supporting open, shared and collaborative work-spaces and hubs: Recent transformations and policy implications. *Urban Research and Practice, 13*(3), 348–357. https://doi.org/10.1080/17535069.2019.1674501

Berardi, F. (2015). *The soul at work: From alienation to autonomy* (F. Cadel & G. Mecchia, Trans.). Semiotext(e).

Berlant, L. (2016). The commons: Infrastructures for troubling times. *Environment and Planning D: Society and Space, 34*(3), 393–419. https://doi.org/10.1177/0263775816645989

Bollier, D. (2008). *Viral spiral: How the commoners built a digital republic of their own*. New Press.

Boltanski, L., & Chiapello, È. (2007). *The new spirit of capitalism* (G. Elliott, Trans.). Verso.

Buden, B. (2007). *The post-Yugoslavian condition of institutional critique: An introduction on critique as countercultural translation*. Transversal. European Institute for Progressive Cultural Politics. Available at: https://transversal.at/transversal/0208/buden/en

Caffentzis, G., & Federici, S. (2014). Commons against and beyond capitalism. *Community Development Journal, 49*(1), 92–105. Available at: https://www.jstor.org/stable/26166211

Child, D., Reckitt, H., & Richards, J. (2017). Labours of love: A conversation on art, gender and social reproduction. *Third Text, 31*(1), 147–168. https://doi.org/10.1080/09528822.2017.1365492

Čukić, I., Timotijević, J., Stojić, B., Dragusha, N., Rexha, O., Dragović, S., & Rajić, T. (2020). *Spaces of commoning: Urban commons in the ex-Yu region*. Belgrade, Ministry of Space: Institute for Urban Politics. Available at: https://rs.boell.org/sites/default/files/2020-09/FINAL_web_COMMONS%20ExYU%20-%20spreads.pdf

De Angelis, M. (2017). *Omnia sunt communia: On the commons and the transformation to postcapitalism*. Zed Books.

De Ruyter, A., & Brown, M. (2019). *The gig economy*. Agenda Publishing.

Deleuze, G., & Guattari, F. (2004). *A thousand plateaus: Capitalism and schizophrenia*. Continuum. (Original work published 1987).

Dević, A. (2021). Bones, muscles, and connective tissue: Tales of collectivity. *E-flux, 119*. Available at: http://worker01.e-flux.com/pdf/article_400856.pdf

Federici, S. (2004). *Caliban and the witch: Women, the body and primitive accumulation*. Autonomedia.

Fisher, M. (2009). *Capitalist realism; Is there no alternative*. Zero Books.

Fisher, M. (2014). *Ghosts of my life: Writings on depression, hauntology and lost futures*. Zero Books.

Frayne, D. (2015). *The refusal of work: The theory and practice of resistance to work*. Zed Books.

Gregg, M., & Seigworth, G. J. (Eds.). (2010). *The affect theory reader*. Duke University Press.

Gržinić, M. (2008). *Re-politicizing art, theory, representation and new media technology*. Vienna, Academy of Fine Arts.

Hardin, G. (1968). The tragedy of the commons. *Science, 162*(3859), 1243–1248. Available at: https://www.jstor.org/stable/1724745

Hardt, M., & Negri, A. (2001). *Empire*. Harvard University Press.

Hardt, M., & Negri, A. (2009). *Commonwealth*. Belknap Press of Harvard University Press.

Harvey, D. (2007). *A brief history of neoliberalism*. Oxford University Press.

Jakovljević, B. (2016). *Alienation effects: Performance and self-management in Yugoslavia, 1945–91*. University of Michigan Press.

Kirn, G. (2019). *Partisan ruptures: Self-management, market reform and the spectre of socialist Yugoslavia*. Pluto Press.

Lazzarato, M. (1996). Immaterial labour. In M. Hardt & P. Virno (Eds.), *Radical thought in Italy: A potential politics*. University of Minnesota Press.

Linebaugh, P. (2008). *The magna carta manifesto: Liberties and commons for all.* University of California Press.
Magacin. (n.d.). *The model.* Available at: https://kcmagacin.org/en/the-model/
McRobbie, A. (2002). Clubs to companies: Notes on the decline of political culture in speeded up creative worlds. *Cultural Studies, 16*(4), 516–531. https://doi.org/10.1080/09502380210139098
Mišković, D. (2018). Rojc community centre in Pula. In D. Vidović (Ed.), *Do it together: Practices and tendencies of participatory governance in culture in the Republic of Croatia.* Kultura Nove. Available at: https://participatory-governance-in-culture.net/uploads/biljeske_EN_web.pdf
Mitropoulos, A. (2012). *Contract and contagion: From biopolitics to oikonomia.* Autonomedia.
Mouffe, C. (2013). *Agonistics: Thinking the world politically.* Verso.
O'Sullivan, S. (2014). *The missing subject of accelerationism.* Mute. Available at: https://www.metamute.org/editorial/articles/missing-subject-accelerationism
Ostrom, E. (1990). Governing the Commons: The Evolution of Institutions for Collective Actions. CambridgeUniversity Press.
Perelman, M. (2001). *The invention of capitalism: Classical political economy and the secret of primitive accumulation.* Duke University Press.
Pusca, A. (2007). Shock, therapy, and post-communist transitions. *Alternatives, 32*, 341–360.
Spade, D. (2020). *Mutual aid: Building solidarity during this crisis (and the next).* Verso.
Spinuzzi, C. (2012). Working alone together: Coworking as emergent collaborative activity. *Journal of Business and Technical Communication, 26*(4), 399–441. https://doi.org/10.1177/1050651912444070
Srnicek, N. (2017). *Platform capitalism.* Wiley.
Stavrides, S. (2019). *Common spaces of urban emancipation.* Manchester University Press.
Szreder, K. (2021). *The ABC of the projectariat: Living and working in a precarious art world.* Manchester University Press.
Terranova, T. (2004). *Network culture: Politics for the information age.* Pluto Press.
The Care Collective. (2020). *The care manifesto: Politics of interdependence.* Verso.
Todorova, M. (1990). Myth-making in European family history: Zadruga revisited. *East European Politics and Society, 4*(1), 30–76. https://doi.org/10.1177/0888325490004001O
Tomašević, T., Horvat, V., Midžić, A., Dragšić, I., Dakić, M. (2018). *Commons in South East Europe: Case of Croatia, Bosnia & Herzegovina and Macedonia.* Zagreb, Institute for Political Ecology. Available at: http://ipe.hr/wp-content/uploads/2018/04/IPE_COMMONS_ENG_web.pdf
Virno, P. (1996). Virtuosity and revolution: The political theory of exodus. In M. Hardt & P. Virno (Eds.), *Radical Thought in Italy: A Potential Politics.* University of Minnesota Press.

Organisational Decline and the Failure in Alternative Organising: The Case of a Coworking Cooperative

Eeva Houtbeckers and Maja Korica

Introduction

> [C]apitalism as we know it has come to an end. (...) We need to construct a different way of developing our societies.
> CEO's greeting, The Community newsletter 2010

Much existing research has focused on examining the formation of alternative organising, the hopeful auspices of largely hopeful efforts. In comparison, the decline of established organisations, including alternative ones, remains understudied (Cederstrom & Fleming, 2012). To address this, we introduce the results of a longitudinal ethnography of a Finnish coworking space 'The Community' established in the late 2000s. We explore how its volunteers, managers, and users related to The Community's gradual decline and finally bankruptcy, and what broader lessons we might take from this.

Specifically, as articulated in the opening quote, many believed that The Community, a coworking cooperative for freelancers, activists and small business owner-managers, would embody their values attached to novel ways of working, which would in turn 'change the world'. This vision of alternative organising was manifested in various processes, including consensus-driven decision-making and blurring of boundaries between roles. The Community thus relates to a larger discussion regarding how we might do things differently in the times of contemporary capitalism (Gibson-Graham, 2006; Parker et al., 2008). While we are increasingly observing the spread of alternatives such as cooperatives, the B Corporation, and

E. Houtbeckers (✉)
Department of Design, Aalto University, Helsinki, Finland
e-mail: eeva.houtbeckers@utu.fi

M. Korica
Management & Society Department, IESEG School of Management, Paris, France
e-mail: m.korica@ieseg.fr

J. Merkel et al. (eds.), *Coworking Spaces*, https://doi.org/10.1007/978-3-031-42268-3_4

social enterprises, empirical studies that explore potential contradictions underpinning efforts at alternative organising remain relatively rare (Fleming & Mandarini, 2011). We thus ask: How did The Community decline and what broader lessons can we take to understand the dynamics of organisational decline in alternative organising?

Part of the empirical answer lies in the very nature of The Community. While coworking spaces are often discussed as enabling new work arrangements (Garrett et al., 2017), this ethnographic study reveals that The Community was not free from distinct contradictions, which undermined its emancipatory potential. Despite its organising principles, its aims to maintain a space of value-based alterity were based on broadly similar working practices to those that they originally aimed to challenge. Although people in The Community believed it to be different, its everyday work practices became similar to mainstream templates it critiqued.

We demonstrate this in three themes that echoed critical steps in the Community's decline: (i) The Community way of doing things (defining the utopic alternative); (ii) working in the space (mundane maintenance); and (iii) working conditions (emergence of functional imperatives). Particularly, we show how relying on managerial and leadership practices developed for commercial corporations also brought in such organisational practices that metaphorically brought the whole house down.

Organisational Decline in Alternative Organising

Organisational decline is one of the many concepts used for expressing organisational unravelling, alongside failure, bankruptcy, and dissolution. They relate to the process of organisations ceasing to exist, against the original wishes of the actors involved. Wilkinson and Mellahi (2005) distinguish between deterministic and voluntaristic forms of organisational decline. In deterministic perspectives, decline is explained by environmental factors that the organisational actors cannot impact, whereas in voluntaristic perspectives decline is explained in terms of psychological and other aspects within the organisation.

There are notable research gaps in the broader literature on organisational decline in three distinct ways. First, as stressed by Josefy et al. (2017), the literature remains methodologically dominated by event history analyses, experiments and modelling, with a dearth of distinct case studies. This results in investigations of environmental contributors or easily measured internal factors, such as public records or founder entrepreneurial experience, at the expense of rich qualitative accounts of intra-organisational dynamics. Second, although a growing number of process-oriented studies explore the mechanisms contributing to decline, *conflict* remains conceptually under-explored relative to the more familiar explanatory focus on rigidity (Habersang et al., 2019). As Habersang et al. (2019) stress, 'examining the systematic interaction of mutually reinforcing mechanisms creating their seemingly opposite effect' holds notable analytical promise, particularly in rarely explored organisational contexts, such as coworking. Third, much existing focus is on incorporated firms, particularly large-scale public entities. There is less attention paid to more temporary forms (Bakker et al., 2016), including coworking cooperatives. As Josefy et al. (2017) note, 'given the potential increase in these forms of organising economic activity, we would encourage future scholarly work in this areas'.

Alongside failure in firms, there are similarly few studies of organisational decline of alternative organising and coworking. Alternative organising has become synonymous with doing things differently, since it is considered to indicate being more aware of, if not avoiding, neoliberal capitalist practices said to increasingly govern and colonise individuals, communities, and non-humans (Gibson-Graham, 2006). Because of this, the organisational decline literature built on commercial companies will likely be less suitable. As Reedy (2014) stressed, alternative organising should not be discussed in terms of what it is not, as it relates to organising as customarily understood in organisation studies, but what it is in itself.

Recent studies on coworking can clarify the conflicts and failures when organising economic activities beyond capitalist understandings. Blagoev et al. (2019) refer to the literature on *organizationality*, which focuses on the interplay between the formal and informal aspects of organising. Their work reveals how coworking spaces have unique layers of these aspects, which can reveal new perspectives to the study of coworking spaces. Furthermore, this complicates the image of a possible failure of a coworking space since there was no one formal organisation to begin with. Moreover, Vidaillet and Bousalham (2020) show how in coworking many organisational realities can take place at the same time. Their findings from three coworking spaces indicate how potentially opposing practices may coexist. The nature of involvement, relationships, and exchanges may include varying practices manifested by the coworkers in a same space. Thus, coworking spaces can be understood as syntopias, possible places for alternative economic organising. Yet according to Vidaillet and Bousalham (2020, p. 83), it is the task of the coworkers to fulfil the potential.

Data and Methodology

This study is based on a longitudinal ethnography conducted during 2009–2015 by the first author. It was part of a larger study focusing on the everyday work within emerging Finnish social enterprises. The Community (pseudonym), a cooperative established in 2009 in a large Finnish city, was a member of a global network of similar coworking spaces. While there were other coworking spaces in Finland, The Community was a pioneer that aimed at 'an open innovation platform where people would feel good, meet each other and do interesting stuff' across sectors (founder-manager, interview 2011). It consisted of cooperative owners (~30), members (i.e., customers, ~80), managers (i.e., cooperative executive board members, ~4 at a time, and the CEO, 3 in total), and hosts (6–12 at a time). Interestingly, actors combined roles, such as founder-investor-manager or owner-member. Moreover, open events organised in the space brought in interested and versatile visitors.

Research Process

The first author started to follow The Community in 2009, when it was a proto-space. In 2011, when the cooperative finally found a suitable space to rent, the first author started to spend time there to shadow the hosts (Czarniawska, 2007), attend events, collect documents, and conduct interviews. A total of 13 interviews were conducted before and after the bankruptcy in 2014 (Table 1).

In total, they spent 31 days doing participant observation and taking fieldnotes. The attention was first on how The Community worked, but later on how the actors managed the tensions emerging from the financial troubles, as a notable emergent category of analytical interest. In addition, they gathered and analysed documents, including internal guidelines. Most interviews were recorded, or notes were written immediately after.

Table 1 Overview of empirical material

Year	Empirical material (no / days)	Interviewees / Location
2009	Observation (2)	The proto space
2010	Interview (1)	The first CEO
	Observation (2)	The proto space
2011	Interviews (2)	Host Head of the executive board
	Observation (14)	A conference organised by the platform (2) The space (12)
2012	Interviews (6)	Cooperative member Head of the executive board Member of the executive board Hosts (2) The first resigned CEO
	Observation (11)	The space (11)
2013	Interviews (4)	Head of the executive board Hosts (2) Resigned head of the executive board
	Observation (4)	A conference organised by the platform (1) The space (3)
2014	Interview (1)	Head of the executive board
2015	Interview (1)	A member of the executive board, host
2010–2014	Newsletters (180)	Publicly delivered e-mails from the community
2010–2015	Internal documents (44/190 pages)	Guidelines, brochures, invitations, memos
2011–2013	Photos (38)	During observation

Analytical Process

The analysis of empirical material was done in collaboration between the two co-authors, following established principles of interpretive research, particularly abductive reasoning (Golden-Biddle & Locke, 2006; Locke et al., 2015). We stayed close to the multiple tensions of everyday organisational interactions and asked questions such as: What was the nature of the aspirational talk between actors? What were the humour and/or sarcasm directed toward? What were the tensions and contradictions? We did these while reading widely across the academic literatures, including organisational decline.

After thematically categorising the material, we compared the aspirational sayings and doings to the tensions and contradictions (Table 2). The resulting analysis presented here thus centres on The Community's 'atmosphere', namely how two components, (1) the mundanity of organising an alternative organisation and (2) doing this in a capitalistic system, together contributed to the gradual erosion of The Community. We outlined several contradictions that highlighted the hard work of maintaining a coworking space in relation to the organisation's aspirational talk. These instances showed how the rise of functional imperatives slowly gained momentum despite opposite aims. This process intensified as The Community's financial situation became increasingly problematic. Outlining these tensions demonstrates how the same features of organising maintained The Community *and* resulted in its decline.

The Decline of The Community

Element I: The Community Way

> I often describe our platform as this kind of a state of mind. So we are often asked when we talk about a collective co-working space that people say "oh, you are an office hotel." So we do have elements from an office hotel, elements from a member club, elements from a think-tank-type of a thing. Although I think we are more of a do-tank, action oriented. But The Community concept is very unique. We often tell people that come and visit our platform because then it [the atmosphere] verifies the best (1st CEO, interview 2010).

Throughout its existence, even the most experienced managers with marketing responsibilities had difficulties in defining The Community. Depending on who was asked, the aspirational talk differed. According to the first CEO interviewed in 2010, the 'vision 2012' was to be a recognised 'change catalyst'. A way in which its alterity was seemingly enacted was via *structures of membership*. While initially The Community was established as a cooperative, actors purposefully blurred the roles between cooperative members, space users, visitors, volunteers, and managers. There was a sense of excitement and possibility. The inspiring working space attracted professionals from existing sectors, like graphic design, and later from

Table 2 Elements and themes from analysis reflecting the porosity of The Community

Elements	Themes	Constructive instances	Destructive instances
The community way of doing things	Experiencing an open purpose	It is hard to describe what The Community is. Instead, one has to experience it (2012) Two cooperative members: The Community is 'more than space' *it is* 'a community' (Coop meeting, 2012)	Member-user-funder: We need to decide what we communicate, what we are. When we know the answer, the communication gets easier (2012) Visitor: 'Many enterprises fail because the entrepreneur attends free events instead of doing hard and boring work. The balance is important'. Host: 'The Community is full of temptations' (2011)
	Emphasising good things	The resigned head of the executive board: 'There is a need for actors like The Community who do things and not just produce paper' (2013)	The head of the executive board: 'People come and say it is a good thing. Sure, it is. But where is the income/money? Whatever good idea you have, you have to find money for it. Or do it yourself for free' (Cooperative meeting, 2012)
	Linking diverse people, projects, and perspectives	'Where the unlikely allies meet' (the first CEO, interview December 2010) Member: 'I want to get to know others and generate Community spirit' (Internal meeting, 2011) Member who had experienced bullying at work before stated 1 year after being in The Community: 'I feel much better' (Observation 2012)	Being active means also contributing to The Community: 'I wouldn't want to see The Community as a regular working space […] Every member has to be ready to take The Community forward. […] Let's find ways how to sparkle one another, share ideas, develop' (User, opening event, 2011) 'We don't ask them [cooperative members] anything, they just pay' (Host, 2012) Approximately 10% of members actually use the space (Opening event, 2011)
	Focusing on now and the future	A sense of unhurriedness for carrying out tasks: Guiding interns, sorting out expenses, going to the shop, calmly talking to people (2011) Only 3 h after a breakfast meeting, all the people have finally left or continued their work (2011) A new member of the executive board: 'Sooner or later The Communities have managed to succeed elsewhere. This will bring in income for the people involved' (2013)	Two hosts teach the new intern how to make coffee while people are coming and going, and some could be approached by one of the hosts (2011) The new hosts seem uninterested in the ethnographer's knowledge concerning issues from 1 year ago (2012)

Elements	Themes	Constructive instances	Destructive instances
Working in the space	Undefined hosting	The friendly and warm behaviour of hosts makes one feel welcomed: Offering coffee, asking questions, smiling (Internal hosting manual, 2011) Hosting enables connections: 'That's why we, hosts, are here for. To answer your questions' (Host, 2011)	Hosts (mostly young students) tend to force members (users) into specific ways of being. For instance, it should be possible to talk about work related projects during The Community After Work on Fridays (2012)
	Open decision-making	Democratic task division in and between meetings (Observation 2011, 2012) The Community described as 'DIY space' in the members' orientation (2011) Member-user-funder: The Community is like a buffet, possibilities for many things (2012) 'Things have changed in a year' (A new host, 2012)	Decision-making is unclear: One 'asks around' and then 'someone decides' (2011) Little agreed means for internal processes (2011) or they change dramatically in just few weeks (Observation 2012, 2013) 'Not much has changed' (An old host, 2012) There are many tasks to cover in a one-hour internal meeting. The head of the executive board seems to want to carry on efficiently but two hosts go against it by problematising issues (2011)
	Transcending financial responsibility	Trying out (at the time) a new method of finance, peer-to-peer loan (CEO, interview 2010)	'So actually [I do] all kinds of things. Except finances' (Host, interview 2013)
Working conditions	Relying on volunteering	The head of the executive board concludes that the atmosphere has not been ruined by the financial problems, quite contrary, people have been working together. A member of the cooperative concludes that a lot is happening in The Community (Coop meeting, 2012) After CEO's burn out 'we [hosts] are afraid that the responsible manager gets ill … but we are like a family so now we try harder, the host group has become tighter … we wish that the CEO gets well' (Host, 2011) Seminar series by The Community members, managers and hosts to bring in income for The Community after the revelation of financial troubles (Observation 2012)	Payments for two actives from The Community can be postponed as they are in the executive board (Coop meeting, 2012) A small number of volunteers (hosts and managers) serve entrepreneurs and freelancers who pay very little for the space (Responsible manager, interview 2012)
	Little limitations to carry on with versatile projects	Hosts using the space to work on their own projects (Observation & interviews 2011, 2012)	'Crappy jobs' (Observation 2011) CEO burnout (Observation & interviews 2012) One host stated that they limited the time used for The Community (Interview 2012)

emerging ones, like graphic facilitation and crowdfunding. Overall, there was a strong trust in people's ability to create new practices that 'change the world together', as the Wi-Fi password stated.

Despite its apparent diversity, both in conceptions of alterity and membership, the Community did practically identify certain members as 'more ideal', i.e., more suitable. The increasing financial difficulties resulted in discussions regarding how to generate more income. While some thought the answer was getting more cooperative members, others advocated for more users. Particularly, the managers and hosts of the space saw the cooperative members as passive and uncommitted, since most of them did not spend time in The Community, nor invested more money when asked. This contrasted with the users, who were present in the space more frequently and paid for the space. When discussing how to get ideas for improved decision making, one host joked: 'We don't ask them [cooperative members], they just pay'.

Particularly, The Community was marketed as 'the buzzing place to be'. Yet, during the observations, it came apparent that there were times with very few people attending events or working in the space. In addition, the aim was to make it a place for social entrepreneurs. Indeed, 'The Community way of doing things' was referred to in interactions, meaning openness in purpose, and activities with the aim of supporting 'good things'. Good things here referred to any project the users of the space considered worthwhile, such as using The Community space for free for various users' social events. Yet, there were no apparent activist projects after the move to the newly renovated space. Instead, paying customers were the focus of member attraction in the new space and links to corporations, like a global ICT company, were nurtured.

Element II: The Work of Mundane Maintenance

> It would be easier to ask what I have not done. [laughs] [...] But in practice I have run this platform like here in everyday routine. Dealt with everything, everything possible from space bookings and space contracts, from cleaning to like everything. [...] Except finances (Host, interview 2013).

The work of The Community as the envisioned alternative was most notably accomplished via *hosting*, which displayed key features of the eventual emerging contradictions, challenging the work of community maintenance. Specifically, hosting created a homey environment. The non-work atmosphere was in fact so successful that one member decided to leave the space because it was 'too much fun' (Manager, interview 2012). Yet, despite this lofty focus on connecting people, it was acknowledged in everyday talk that 'crappy jobs' nevertheless piled up. According to one host, they 'nurtured' members. Hosts took time to solve mundane issues that might have seemed small, like organising the fixing a broken microwave oven, but in fact were crucial for the success of the space. Despite the hosting manual's warnings, hosting bared an aspect of 'servitude' not considered lucrative for people with a more stable job. Despite this, the managers underestimated the effort needed for nurturing daily tasks of hosts. One manager commented the work of building boxes for users to store their things as 'tick, tock, tock and they are done' (interview 2011). They seemed to have little idea what it would require to first get enough people together and then actually carry out the task.

In late 2011, financial problems emerged, causing critical evaluation of many practices. The prioritisation of cash flow to save The Community created an even greater division of tasks between managers, responsible for the finances, and hosts, responsible for maintaining everyday activities. Soon, there were different crappy jobs depending on one's position in the Community. While hosts bore a lot of emotional baggage and acted in the interface between the members and the official decision-makers, the hosts categorically dismissed themselves from 'finances'. Finances were a matter for the (mostly non-paid) managers. This was reinforced by the decision-making structures, according to which the changing CEOs and the active management formed the highest financial authority. The division of labour also accentuated the undervaluation of others' work: the managers were unable to acknowledge the time needed for the mundane work of hosting, while the hosts were unable to acknowledge managers' constant pressures related to securing the cash flow.

Element III: Working Conditions and the Rise of (Dys) functional Imperatives

The Community required people to work for it. But due to its labour-intensive processes and scarce financial resources, it was dependent on unpaid work of volunteers. Indeed, only a handful of people were paid. This was explained by the start-up status and financial realities, but it was also absorbed to 'The Community way'. However, the management of volunteer unpaid labour was problematic. Hosts left or reduced their work input at short notice. The managers were less likely to leave, but this was because they had more financial connection, in the form of paid work or own financial investments, or official responsibilities due to their status. As a result, The Community made 'use of the efforts and innovations among groups and individuals' beyond formally employed, i.e., un(der)paid volunteer work (Fleming, 2014: 885).

Such economic valorisation increased after the revelation of the financial troubles. As an urgent measure, they established a high-profile seminar series produced by the users of the space without compensation. Another example was hosts' work, which early managers considered to require special skills and as an important element of The Community. Although there were plans to stabilise hosts' position by employment, there were no resources. As if this was not already a compromise, the demands increased, thus further normalising overwork. One host described the way they had deliberately tried to limit their work time, since this was 'good for it [The Community] but not sensible for my own finances' (interview 2012).

When the financial troubles accumulated, a new manager further advocated separating The Community activities into a 'social club' and a 'business enterprise'. They felt the hosts were not acting efficiently enough when running daily tasks. Instead of challenging this binary thinking, their solution was to emphasise 'conventional work' (Fleming & Mandarini, 2011, pp. 339–340), which they referred to as 'business enterprise', and abolish 'social club', the very thing that kept The Community together.

In early 2012, the first CEO resigned because they felt they had to carry the workload of low income and high expenses alone. Before their burnout, they described not sleeping well and worrying about The Community (interview 2011), despite 'fine moments'. They also normalised the experience, by stating that 'the burnout here is nothing unique' when looking at the global network of The Communities. They admitted they had founded The Community with scarce plans and that they would have needed a separate team for running it.

In summer 2011 at the house-warming party, the contradiction between sayings and doings became unsustainable. One of the founder-managers told the history of the cooperative and indicated that soon after the start they realised the need to hire someone, but it was challenging to find someone who would work "60 hours per week'. Later in the same event, the hired CEO commented that one of the core elements of The Community is 'well-being' and 'everything else is just a means for this goal'—while they had just returned to work after 2 months of sick leave. Thus, the high workload customarily present in non-alternative corporate settings appeared here as well and was institutionally accepted, with little recognition of the conflict with the stated aims of The Community.

Despite exciting activities of The Community, in 2012 it came apparent that it had not achieved its 'Vision 2012'. It limped due to its passive owners (cooperative members) and high loans. While The Community was barely able to maintain the space by subleasing it to users, they were out of resources to develop collective activities. Moreover, managers changed often, and regardless of the commitment among hosts, their turnover was similarly noticeable, with considerable periods of absence.

While invested The Community actors tried to change the status quo and enable resources for more paid work, financial difficulties prevailed. The reasons for the unhealthy financial situation were clear. Income was generated from memberships, projects and selling the space. Yet, there was the burden of high renovation expenses, low sales of new memberships, and unrealised project funding. The second CEO commented that these were the front-heavy expenses like in any start-up. There was no income to hire more people although there were more members. The Communities across the globe had small margins of revenues while a lot of work was done in them. Although they needed a stable stream of revenue, the pricing was kept low to enable the diversity of users, i.e., small entrepreneurs. This darker sides of volunteer work were brought up by a new manager in an interview in 2012. They commented that a small number of volunteers, hosts and managers, served profit-seeking entrepreneurs, who paid very little for their coworking space.

To balance the finances after the crisis in late 2011 and early 2012, management discontinued unprofitable projects and memberships, leading to more business-like activities. One host complained there was 'too much focus on money' (interview 2012). But for the managers, who had invested considerable sums, this perspective seemed absurd. Amid the financial crisis, rescue attempts were thus developed:

> The idea was to sell [The Community] spirit in a bottle to companies. So air in a glass bottle. I was strongly against it because I think selling such crap is not part of this. But there were many conservative representatives and older gentlemen who said that sure, companies buy whatever vases and pay terrible amount of money for them in order to support a good cause (Host, interview 2012).

The last CEO referred to 'The Community spirit' as serendipity, which was still there despite the grave financial troubles, but it was not enough to support it. To them, the downside of serendipity was *'unskilful [...] naivety [...] a belief in what was nice to believe in but it isn't real'*. Finally, the Community was formally no more.

Discussion: The Community's Emancipatory and Transformative Potential

Despite The Community's problems, its bankruptcy was a surprise to the ethnographer (Houtbeckers, 2017). While the aspirational talk and mundane practices suggested that the people involved would push further despite the challenges, the financial debt together with the organisational arrangements led to lack of means and will. This raises questions on what could have been done differently to ensure both the bottom-up character and the sustainability of the Community and other cases of alternative organising. What kind of organisational arrangements could have provided the Community with a different path?

Hosts' and managers' status as volunteers could have brought them together, alongside owners/members, to solve the issues of unpaid labour and structural injustices and leverage the alternative potential of coworking (Vidaillet & Bousalham, 2020, p. 83). Yet, this did not happen. Instead, the cooperative model was not leveraged to organise the space and related commitments as commons. From early on, the members of the cooperative were different from the users of the space, who acted as the customers rather than members—although they were called members in everyday talk. The cooperative members (owners) in turn rarely visited the space or participated in sustaining 'the Community spirit', despite some of them later seeing it as a critical commodity to sell in an effort to save it. As a result, neither the users nor the cooperative members had a clear incentive to invest effort into developing the space, let alone financial assets needed to cover the debts. The unpaid hosts had even less—their unpaid labour upheld the Community, but without a real stake in its finances or means of participating in its decision-making. As the last (unpaid) CEO commented after the bankruptcy, there was a 'casting defect' in The Community (interview 2014).

Such tensions partly mirror broader labour dynamics in recent decades, including the emergence of a new 'precariat' (Ross, 2009) of individuals (apparently) eager for work experience, who largely work short term and are paid little or none. This can escalate to exploitation, whereby '(a) Y [here volunteer, but also 60 hours a week manager] is forced to make a sacrifice which results in benefit for X [Community owners] and (b) X [here Community owner] obtains this benefit by means of advantage in power that X [it] has over Y [volunteers/unpaid managers]' (Shelby, 2002: 393). This is especially problematic in settings driven by passion or belief in larger mission (Hochschild, 1983; Cech, 2021), where 'passion exploitation' (Kim et al., 2020) can quickly arise in the form of unfair and/or demeaning employment practices, such as asking workers to do menial tasks or not paying them appropriately for extra work. This is sustained by a belief that work in and of itself is sufficient recognition for a passionate individual, i.e., meaningful work, fair pay, a financial stake, and a say in the organisation are optional. This suggests that it is not sufficient for volunteers and members to merely share a belief in the mission itself, sustained by close social interactions in the Community's physical space. If the labour of both is needed to sustain the organisation over time, then this should be appropriately compensated as work and/or formal shared ownership. Indeed, as Parker et al. (2008, p. 61) noted, 'cooperative members contribute to and control the capital of their cooperative', which is generally organised democratically as well. It can thus be argued that The Community's failure to inclusively organise as a cooperative, that is to practice related principles regarding decisions and finances that include all its actors, was a key facet of its eventual demise.

This point relates to another tension that arose from our study of The Community, namely the dangers of celebrated informality, in contrast to 'formality' associated with managerial hierarchy. The Community leaders often praised its 'Spirit' and values activated through volunteer-driven social events, believing that this, rather than organisational plans, structures, and hierarchies, ensured its success. Not only were managers mostly unpaid, despite heavy workloads, plans were largely absent, and hierarchies were fluid, with little formal coordination between hosts or indeed their stability over time. In this, the Community echoed broader movements away from managerialist organisations and toward less-hierarchical forms of organising, driven by increased tendencies to equate work with meaning (Lee & Edmondson, 2017). However, the contradictions we observed also demonstrate that alternatives to hierarchy do not necessarily mean no structures, as evident in the decentralised management practice Holacracy for instance (Robertson, 2015). Indeed, they evidence the importance of appropriate support to self-management efforts, without which critical features like increased importance of coordination cannot be easily sustained. Key to such support is 'well-defined decision rights', including effective 'property rights for all employees' (Lee & Edmondson, 2017); a feature that was notably absent here.

Finally, although this chapter does not allow us to fully elaborate it, The Community also holds potential lessons for understanding how alternative organisations might successfully balance competing institutional logics (Smets et al., 2015), in this case, 'utopian visions' versus 'financial imperatives'. This is especially when they face organisational decline prompting greater emphasis on the latter, making it increasingly difficult to exploit their mutual benefits, or keep those logics co-existing, as happened in our case with the cancelling of the social club in favour of the business club. One argument deserving of further empirical exploration is whether it is precisely the deletion of The Community's normative logic of 'utopian visions' at a time when tensions between the two logics increased which precipitated its failure, as opposed to it exploring other 'balancing mechanisms' that might sustain them both over time (ibid).

Conclusion

While we are sympathetic to the calls for alternative organising, this study has sought to contribute to the lack of rich descriptions of organisational decline in its contemporary settings. The Community as a coworking space had the potential to challenge 'capitalism as we know it', as referred to in the opening quote, in light of its declared alternative organising principles. Yet, the specific contours of its supposed alternative organising also resulted in its decline. There were contradictions arising from the presence of the same valuations plaguing modern work, including unpaid work and burnout, and bottom-line considerations trumping everything else.

This chapter developed the detailed understanding of contradictions in coworking (Vidaillet & Bousalham, 2020; Blagoev et al., 2019). We depicted empirically the mundane dynamics and contradictions of a coworking space. By describing the working conditions and the disagreements on the direction of The Community, the chapter richly illustrated the underlying conflicts that arise in organising and demonstrated that alternative organising requires hard work too. Thus, the chapter also holds important practical implications. Particularly, it addresses the important question of how we can and do organise spaces that enable alterity, especially those that envisage and organise passion as a liberatory principle (Fourier, 1968).

Specifically, the last decades have been characterised by a normatively led search for 'better', including alternative organisational forms such as cooperatives, social enterprises or coworking. One of the most burning questions in alternative organising is how some initiatives succeed and some fail, in part due to wider, often opportunistic depictions of past utopian experiments, like early twentieth century Californian cooperative settlement New Llano, as inevitable failures (see Parker et al., 2008). Yet pragmatic questions like 'how do we keep going to make this work' remain relatively underexplored. This is surprising least of all because most new entrepreneurial initiatives fail (Welter et al., 2017), not just those driven by cooperative principles. As such, by explicating the mechanisms of one relevant account of decline, this study may be directly useful for practitioners working daily to sustain such alternative spaces. Moreover, by highlighting the processes that constructed but also crumbled alterity, practitioners may gain further appreciation of the contemporary double-edged sword of such spaces. In other words, we suggest there continues to be much to learn from organisational decline, as from our more customarily explored tales of success.

References

Bakker, R. M., DeFillippi, R. J., Schwab, A., & Sydow, J. (2016). Temporary organizing: Promises, processes, problems. *Organization Studies, 37*(12), 1703–1719. https://doi.org/10.1177/0170840616655982

Blagoev, B., Costas, J., & Kärreman, D. (2019). 'We are all herd animals': Community and organizationality in coworking spaces. *Organization, 26*(6), 894–916. https://doi.org/10.1177/1350508418821008

Cech, E. (2021). *The trouble with passion: How searching for fulfilment at work fosters inequality*. University of California Press.

Cederstrom, C., & Fleming, P. (2012). *Dead man working*. John Hunt Publishing.

Czarniawska, B. (2007). *Shadowing: And other techniques for doing fieldwork in modern societies*. Copenhagen Business School Press.

Fleming, P. (2014). Review article: When "life itself" goes to work: Reviewing shifts in organizational life through the lens of biopower. *Human Relations, 67*(7), 875–901. https://doi.org/10.1177/0018726713508142

Fleming, P., & Mandarini, M. (2011). Towards a workers' society? New perspectives on work and emancipation. In M. Alvesson, T. Bridgam, & H. Willmott (Eds.), *The Oxford handbook of critical management studies* (pp. 328–344). Oxford Handbooks.

Fourier, C. (1968 [1851]). *The passions of the human soul and their influence on society and civilization*. Kelley.

Garrett, L. E., Spreitzer, G. M., & Bacevice, P. A. (2017). Co-constructing a sense of community at work: The emergence of community in coworking spaces. *Organization Studies, 38*(6), 821–842. https://doi.org/10.1177/0170840616685354

Gibson-Graham, J.-K. (2006 [1996]). *The end of capitalism (as we knew it): A feminist critique of political economy* (2nd ed.) University of Minnesota Press.

Golden-Biddle, K., & Locke, K. (2006). *Composing qualitative research*. Sage.

Habersang, S., Küberling-Jost, J., Reihlen, M., & Seckler, C. (2019). A process perspective on organizational failure: A qualitative meta-analysis. *Journal of Management Studies, 56*(1), 19–56. https://doi.org/10.1111/joms.12341

Hochschild, A. R. (1983). *The managed heart: The commercialisation of human feeling*. The University of California Press.

Houtbeckers, E. (2017). Researcher subjectivity in social entrepreneurship ethnographies: The entanglement of stories in a co-working cooperative for social innovation. *Social Enterprise Journal, 13*(2), 128–143. https://doi.org/10.1108/SEJ-07-2016-0025

Josefy, M. A., Harrison, J. S., Sirmon, D. G., & Carnes, C. (2017). Living and dying: Synthesizing the literature on firm survival and failure across stages of development. *Academy of Management Annals, 11*(2), 770–799. https://doi.org/10.5465/annals.2015.0148

Kim, J. Y., Campbell, T. H., Shepard, S., & Kay, A. C. (2020). Understanding contemporary forms of exploitation: Attributions of passion serve to legitimize poor treatment of workers. *Journal of Personality and Social Psychology, 118*(1), 121–148. https://doi.org/10.1037/pspi0000190

Lee, M. Y., & Edmondson, A. C. (2017). Self-managing organizations: Exploring the limits of less-hierarchical organizing. *Research in Organizational Behavior, 37*, 35–58. https://doi.org/10.1016/j.riob.2017.10.002

Locke, K., Feldman, M. S., & Golden-Biddle, K. (2015). Discovery, validation and live coding. In K. Elsbach & R. Kramer (Eds.), *The handbook of qualitative organizational research* (pp. 371–380). Taylor and Francis.

Parker, M., Fournier, V., & Reedy, P. (2008). *The dictionary of alternatives: Utopianism and organization*. Bloomsbury Publishing.

Reedy, P. (2014). Impossible organisations: Anarchism and organisational praxis. *Ephemera Theory & Politics in Organization, 14*(4), 639–658.

Robertson, B. J. (2015). Holacracy: The new management system for a rapidly changing world. Henry Holt and Company.

Ross, A. (2009). *Nice work if you can get it*. New York University Press.

Shelby, T. (2002). Parasites, pimps and capitalists: A natural conception exploitation. *Social Theory and Practice, 28*(3), 341–418. https://www.jstor.org/stable/23562053

Smets, M., Jarzabkowski, P., Burke, G., & Spee, P. (2015). Reinsurance trading in Lloyd's of London: Balancing conflicting-yet-complementary logics in practice. *Academy of Management Journal, 58*(3), 932–970. https://doi.org/10.5465/amj.2012.0638

Vidaillet, B., & Bousalham, Y. (2020). Coworking spaces as places where economic diversity can be articulated: Towards a theory of syntopia. *Organization, 27*(1), 60–87. https://doi.org/10.1177/1350508418794003

Welter, F., Baker, T., Audretsch, D. B., & Gartner, W. B. (2017). Everyday entrepreneurship—A call for entrepreneurship research to embrace entrepreneurial diversity. *Entrepreneurship Theory and Practice, 41*(3), 311–321. https://doi.org/10.1111/etap.12258

Wilkinson, A., & Mellahi, K. (2005). Organizational failure. *Long Range Planning, 38*(3), 233–238.

CWS as Terrains for the (De)territorialisation of Labour: Digital Nomads, Local Coworkers and the Pursuit for Resilient and Cooperative Spaces

Dimitris Pettas and Vasilis Avdikos

Introduction: Changing Labour Landscapes and Ongoing Challenges for CWS

Over the last decades, under the pressure of developments related to globalisation and digital transformation and, more recently, due to the emergence of the platform economy and the implications of the COVID-19 pandemic, processes of production, distribution and consumption are rapidly changing on a global level. New economic contexts are characterised by increased mobility and connectivity, including flows of people, finance, information and the overcoming of local/global divisions and spatially bounded conceptualisations of economic activities, in favour of more fluid understanding of the latter in the form of networks and circuits that connect distant and differentiated sites (Anderson & McFarlane, 2011; Anderson et al., 2012; Law, 2004; Murdoch, 1998; Simandan, 2017; Collinge, 2006), also through the rapidly increasing mediation of digital environments. The rise and expansion of the platform economy has also resulted in the continuous and increasing splitting of value chains, where large companies find it more profitable to outsource specific parts of their supply chains to third parties, and especially to freelance labour.

In this context, labour is transforming, as platform-mediated and remote modes of labour are gradually established as a widespread normality, also beyond workers whose line of work is directly linked to ICT-related activities. The expansion of remote work during the COVID-19 pandemic is expected to endure and further expand in the post-pandemic period (Eurofound, 2020; Bernstein et al., 2020),

D. Pettas
Institute of Urban and Regional Planning, Technische Universität Berlin, Berlin, Germany
e-mail: pettasdimit@fmail.com

V. Avdikos (✉)
Department of Economic and Regional Development, Panteion University, Athens, Greece

J. Merkel et al. (eds.), *Coworking Spaces*, https://doi.org/10.1007/978-3-031-42268-3_5

especially concerning high skilled and educated labour force (Gaudecker et al., 2020). This trend already enables middle-class remote workers, framed as 'digital nomads' and 'lifestyle migrants' to relocate short or mid-term to cities in which they can exercise their economic superiority, as 'flexible labour and remote working policies are tuned to middle-class aspirations of freedom that nurture spatial inquiries for consumption' (Alexandri & Janoschka, 2020: 3207). Digital nomads blend tourism, leisure, and labour to satisfy their desired work while experimenting with different cultures in different cities and socio-economic fabrics. Moreover, aside from a low cost of living, digital nomads select destinations based on their potential for tourism-related and self-development activities (Bonneau & Aroles, 2021).

We argue that a number of CWS, through their operation as terrains for the territorialisation of workers, are rapidly adapting to the aforementioned developments, while these adaptations i) undermine CWS' capacities and potentialities to constitute alternative and resilient labour arrangements that empower workers and ii) can result in putting pressure in Southern European cities, which are already struggling with the implications of overtourism and transnational gentrification (see Pettas et al., 2022; Drăgan & Camară, 2021). CWS, in their early stages, largely operated as alternative labour arrangements for precarious freelancers, originating mostly from the cultural and creative industries (Merkel, 2015). Moreover, they were predominantly addressed to local labour forces, namely workers who resided and worked in the cities in which CWS operated. In several cases, these conditions resulted in spatiotemporal arrangements that created the conditions for the development of communities of coworkers, as workers remained in CWS for long periods of time enabling interactions and encounters that could lead to the emergence of stable relations and networks while, at the same time, a substantial share of coworking communities comprised of workers who had strong connections to the urban environments of hosting cities. There are several examples where CWS, either coworking spaces or fab labs, makerspaces, and the like, have contributed to the stabilisation of the turbulent routines of project-based workers and provided a safety net to the precarisation of freelance labour.

Nevertheless, the second and third waves of coworking (Gandini & Cossu, 2021) signified the disassociation of CWS from their initial bottom-up approach and their gradual alignment with market and corporate logics, developments which are largely expressed in the turn to 'coworking as a service'. During the last years, coworking services have been incorporated in spaces beyond 'traditional' CWS, namely hotels, airports, etc. Moreover, CWS, especially in popular destination cities in South Europe (like Portugal or Greece), are experiencing a substantial shift concerning their users' profiles, as digital nomads, and remote workers, commonly comprising of international, short-term visitors are growingly accounting for large shares of the spaces' users (Borges et al., 2022; Avdikos et al., 2022; Bosworth et al., 2021). In this frame, we argue that, beyond services that relate to the provision of working spaces for a fee, CWS are rapidly associated with the tourism sector, re-emerging as actors who provide 'access as a service', while labour is gradually becoming a secondary element of CWS' operation, which is now re-arranged around introducing and channelling short-term users who constitute a hybrid mode of tourist-workers [or half tourists, as Almeida & Belezas (2022) coined them] in the urban environment. More specifically, drawing on data from an extensive fieldwork in Athens, Greece, we argue that CWS', that host digital nomads, newly emerged role comprises of access to networks, resources/infrastructure and information concerning the local socio-economic environment. While such a focus could ease the mobile lifestyles of digital nomads, it may also cause disruptions in the communities of domestic coworkers and, furthermore, exacerbate various socio-spatial inequalities in both the internal environment of CWS and the urban environment of cities in Southern European countries. In that respect, we argue that we need to re-think and re-evaluate the role of the more community-led CWS, their core missions and the ways they could still provide a safe environment for the precarious urban workforce. For the latter, we propose that the alignment of CWS with the principles and logics of the social and solidarity economy and the cooperative movement can be helpful towards meeting that goal.

The Rise and Establishment of Remote Work

Digital nomads and lifestyle migrants are mobilised by their desire to work remotely, while also experiencing different lifestyles and cultures, and prioritising variables such as cheaper living costs and good weather, when it comes to the decision where to relocate (Bozzi, 2020; Hall et al., 2019). According to Mancinelli (2020), digital nomads consider travelling as a self-focused experience and a founding element of identity, through which they attempt to define their lives as an individualistic project of self-realisation.

Recent statistics show that most digital nomads are well educated professionals, mostly in their 30's, working mainly in the IT sector and the CCIs, while two out of five digital nomads are females. Statista (2022) reports that 34% of digital nomads earn between $50,000–$100,000 per year, another 33% earn $100,000–$250,000 a year, while their average income is around $120,000. During the pandemic, the volume of digital nomads has risen substantially, as in 2022 there were about 35 million remotely working professionals worldwide. The MBO Partners '2022 State of Independence' research study found that 16.9 million American workers currently describe themselves as digital nomads, while their rates in the total workforce increased by 9% compared to 2021 and by a staggering 131% compared to the pre-pandemic year of 2019. Popular destinations for digital nomads include countries of the Global South, such as Mexico, Thailand, Indonesia, Colombia while, recently, we have witnessed the emergence of new destinations in Europe like Portugal, Serbia, and Greece. These countries share several common characteristics, such as low living cost (in comparison with Northern European countries), touristic attractions and a Mediterranean/ Balkan lifestyle. On the policy side, during the last few years, many countries have introduced Digital Nomad Visas that provide this new stratum of tourist-workers with notable benefits and rights (from the legal right to work to low taxation and the right to apply for permanent residency after several years, amongst other) when choosing to reside in a country for many months, or even years. Greece introduced the Digital Nomad Visa initiative in October 2022 and in that first month more than 3,000 applications were received, while we estimate that until March 2023, about 16,000 people have applied for such a visa.

CWS seem to host about 20% of the overall population of digital nomads (Statista, 2022). The latter visit CWS and work from there as, apart from the fast internet connection service that CWS usually offer, they also assist digital nomads to socialise with like-minded individuals working in the same industries, while also providing valuable information about the local socio-economic environment of the destination cities. During extensive field work in CWS in Athens (2021) we came across many digital nomads, whereas in some CWS almost half of the coworkers were digital nomads. Although they were always present in CWS in the past few years, their numbers considerably rose during the pandemic, causing national media to talk about a wave of pandemic nomads (Avdikos et al., 2022). The interviews with these new strata of pandemic and digital nomads revealed some interesting facts about the role of CWS. If we consider digital nomadism as a continuous flow of people (and disposable income) in a city, CWS play a pivotal role in territorialising these flows through a number of ways. As CWS offer valuable information to newcomers to Athens, they propose specific amenities to visit, they organise social events in bars and restaurants, as well as visits in galleries and museums. Digital nomads also get information about the local real estate market and where to find the most interesting neighbourhoods with street level culture (Florida, 2002) to reside and socialise. CWS open their social networks to welcoming those new members and they can be seen as tourist guides and as mediators between digital nomads and the city of Athens. All information the digital nomads receive from CWS and their communities of coworkers seems to contribute to a risk-assessment process about most of the risks that are associated with nomadic lifestyle (frequent change of place, isolation, and loneliness, etc.). Collective information exchange in CWS can result in minimising some dangers for the digital nomads when arriving in a new place and can also endow nomads with the ability to adapt more swiftly to changing conditions and risk perceptions, as Ehn et al. (2022) emphasise.

The Role of CWS in Territorialising Digital Nomads and the Impact on the Urban Environment

CWS receive nomadic flows, recodify, and metabolise them in the city, channelling nomadic consumption patterns in specific neighbourhoods, as well as directing them to other touristic towns and islands in Greece that are becoming hubs for digital nomads (e.g. Pelion Mountain and Crete). Thus, CWS act like spatial intersections that internalise through codifications (welcoming, tips about the city, special events and tours, etc.) all these exogenous nomadic flows that aim to explore and exploit the city of Athens. At the same time, digital nomads internationalise Athens' coworking market via enriching the professional (as well as the social) networks of the local CWS and augmenting the buzz of CWS with new information flows. Although during the fieldwork we did not find cases where nomads engaged in formal exchanges with domestic coworkers (such as subcontracting part of a project or initiating new projects), we did find a few informal professional exchanges (such as exchange of soft skills, exchange of information about specific markets, etc).

Moreover, digital nomads' economic superiority (also framed as geographic arbitrage, see Ferriss, 2007; Penney & Dramowicz, 2016) contributes to the ongoing touristification processes of the Athenian city centre (Pettas et al., 2022), as their purchasing parity exceeds by far the average annual income in Greece [16,200 euro (2021)], resulting in further pushes of real estate prices in specific central Athenian districts. This was evident during our fieldwork, as there were numerous examples of nomads who were ready to pay significantly higher rents to secure mid-term accommodation in central, culturally and socially vibrant residential districts of Athens. In most cases, digital nomads choose to reside in districts that, over the last decade and due to the implications of the rapid expansion of the short-term rental market, have struggled with the negative externalities of touristification processes, such as Exarcheia, Koukaki and Pagkrati.

Adaptations Towards Facilitating the Needs of Digital Nomads

All the above signify a new era for the coworking landscape of Athens. Acting like a tourist guide that aims to welcome and ease the arrival of international workers, CWS have developed a number of new services (such as paid guided tours) that expand their initial scope, namely offering affordable shelter to freelancers, start-uppers, and remote workers. As highlighted before, these new services assist and territorialise digital nomads in Athens and Greece. However, these shifts concerning CWS' tenants and services caused a number of disruptions which lead to their deterritorialisation and are affecting domestic coworkers. One of the impacts of the increase of digital nomads' flows in CWS was the introduction of daily passes (e.g. the 'hot desk for drop-in travellers' pass', which was introduced in a CWS in the Athens city centre in early 2023), as most of the CWS did not usually offer such a service before the pandemic. Especially the more community-led CWS (Avdikos & Iliopoulou, 2019) were mainly offering monthly passes. Daily passes may be useful for digital nomads and other remote workers to experience and test the environment of a CWS, before acquiring a monthly pass, but they can also disrupt the flows of people usually found in CWS and more importantly, the various processes and relations between coworkers that compose a community in CWS. Although community building is a buzzword in the coworking literature that entails many kinds of ambivalences (for its actual existence, quality and value, see De Peuter et al., 2017; Spinuzi et al., 2019), the frequent interchange of coworkers' population in a CWS undermines processes of compiling a stable community, along with such spaces' role in facilitating practices of mutual exchanges that could contribute to the confrontation of aspects of precariousness, through practices of collective self-help and self-organisation (Merkel, 2019). Instead, the vibrancy of communities in CWS can be seen as an asset and a commodity that can attract digital nomads.

Thus, in CWS that are developed or transformed towards facilitating the needs of digital nomads, the possibilities for the emergence of actual communities of coworkers are substantially undermined due to a series of reasons. First, such a development leads to major gaps in terms of income, working hours, rights, access to social and health insurance, among groups of underprivileged and privileged coworkers, having on the one hand those who struggle within the harsh labour markets of the Global and European South and, on the other hand, those who have secured streams of income and work according to the conditions in more privileged labour markets of the Global and European North, cashing in on their passport strength, as Thompson notes (2019). Such a development leads to a division of coworkers, as the two groups do not share the same problems, concerns, needs and desires and, thus, the emergence of collective identities is difficult to emerge. At the same time, underprivileged workers can experience additional pressure and anxiety due to the constant comparison with freelancers from the same or similar sectors. On top of that, the latter can also be excluded due to their incapability to afford the increasing access costs. This is evident if we look at the rise of prices of monthly passes during the last few years, as the average price for a hot desk in Athens rose from about 100 euro (2018) to 180 euro (2023), whereas during our fieldwork in several CWS (2021), hot desks were scarce due to the rapid excess of demand, especially on behalf of digital/pandemic nomads.

The inequalities and their various manifestations (increase of prices, different consumption patterns) along with the different desires and aspirations of coworkers cause a continuous deterritorialisation in CWS, illustrated in the new services offered. Still, the focus of CWS remains on providing services that reproduce labour but there are currently some peripheral services that reorient the focus towards the provision of facilities that cater tourists (−workers). Subsequently, CWS increasingly attach to the value chains of the tourism industry to cater for the variegated needs of digital nomads and extract a new kind of value when assembled with the value chain of tourism. At the same time, the tourism industry itself has already begun offering coworking facilities in various ways, such as through co-living arrangements, the introduction of targeted provisions for digital nomads on accommodation platforms (see the relevant section on the Airbnb platform) and the creation of coworking spaces in special designated areas in hotels and airports. Coworking, as seen through the perspective of a new service in a tourism-led economy, can be a valuable component for tourism-driven local growth. This is true for the case of Athens, as Athens has recently emerged as a city break destination, also through city branding that was developed between 2017 and 2020, mainly by the Municipality of Athens, following the successful organisation of the Documenta (2017) exhibition in Athens. Within this period, Athens transformed from a one-day-stop to a year-round city break destination, as 6.3 million tourists visited Athens in 2019 (worlddata.info), as opposed to 2.5 million in 2012 (Kathimerini, 2019). Moreover, the transformation of Athens' touristic identity undertook specific qualitative characteristics, as touristic accommodation infrastructure, facilitated through digitally mediated short-term-rentals, expanded in districts beyond the 'traditionally' touristic ones (Balampanidis et al., 2019). Building on that, we argue that the development of coworking and the emergence of some new coworking side-services (as argued before) should be explored in relation with wider urban processes, and not as isolated phenomena. In the case of Athens, CWS can be seen as another urban amenity for a highly mobile creative class, attracted by the rich cultural heritage of the city and the vibrant contemporary creative economy.

However, besides their potential contribution to local growth, the new spatio-temporal arrangements in CWS that facilitate the needs of digital nomads, render extremely problematic their association with local societies, neighbourhoods, professional and social networks, etc. According to Mancinelli (2020), digital nomads' self-realisation project meets the ideology of entrepreneurialism, allowing them to take advantage of privileged nationalities to navigate the global inequalities of the capitalist system. As it was evident through our fieldwork, short-term coworkers have limited to non-existent knowledge of local conditions and problems, while their temporary staying in cities and neighbourhoods in which CWS are nested also results in the lack of intention to associate with any aspect of the latter, beyond consumption-oriented practices. Following that, the potentialities of CWS to operate as local urban actors that, up to some degree, relate to or try to contribute to local social, spatial, economic, cultural, and political contexts are vastly undermined, along with their chances to partake in the creation of ecosystems of social innovation and solidarity that contribute to social cohesion and development. Moreover, consumption oriented CWS, not only lack the ability to connect with local conditions and problems but, additionally, they accelerate and deepen the negative externalities of touristification processes, especially in cities of the European and Global South, in which privileged digital nomads have significantly higher purchasing power compared to local populations, as seen before. These pressures are expressed and reflected in various aspects of urban life, such as housing and consumption landscapes. CWS operating as agents of touristification, under the influence of broader cultural shifts in tourism that promote the distancing from mass tourism accommodation in favour of 'live like a local' experiences, are channelling short-term visitors to residential areas, causing disruptions to housing markets, as digital nomads are willing and capable of paying significantly higher rents compared to local residents. Moreover, the physical and soft infrastructure for mid-term accommodation is already in place, due to the high densities of dwellings that—over the last years—have been listed in short-term rental platforms such as Airbnb. According to Dagkouli-Kyriakoglou et al. (2022), the infrastructure of short-term rentals (including renovated and equipped apartments, along with the experience and knowledge of hosts) can shift from the short to mid-term accommodation of digital nomads and remote workers. Apart from that, broader negative externalities of touristification can be accelerated by the flows of digital nomads, including the undermining and decline of public infrastructure and services (e.g. schools, public health-care units) due to the diminishing numbers of beneficiaries, the shift of local markets from addressing the everyday needs of permanent residents to facilitating those of short and mid-term residents.

Going back to the destabilisation of CWS in Athens, a common characteristic of these new codifications found in CWS that host digital nomads is the one of spatio-temporality, which can be seen as the core of the deterritorialisation process. Digital nomadism is based on temporal relations, as the nomads usually change destinations in a frequent manner (their staying in one city usually extends from a few weeks to 6–12 months) as their desire and goal is to visit several places while working, either as remote employees or as remote freelancers. Thus, temporality is an esoteric characteristic of nomadism that is brought in the social environment of CWS. Following that, two different kinds of temporal sets of relations are met, and clash in CWS: one that is brought from a nomadic lifestyle and the second brought by project-based work. The latter can be seen as one of the core characteristics of freelance work that contributes to the high degrees of precariousness of freelance labour. The CWS (at least the more community-led ones) usually aim to confront that kind of precariousness by attempting to develop stable communities through the formation of daily/weekly routines (e.g. through community lunches, frequent social events, and training sessions, etc., see Tselepi et al., 2023), but also through actions of solidarity between the coworkers. That stability is now at stake with the continuous flows of nomads that bring another level of spatio-temporality in CWS. Nevertheless, while temporality is a desirable attribute for the nomads, for most freelancers it can be seen as a threat. The difference is that the digital nomads materialise temporality (and individualisation, autonomy, and control) through a superior economic position, that provides them with the opportunity to travel while working, a condition that compensates for the precariousness of remote work. On the contrary, for domestic freelancers/coworkers in Athens, temporality is associated with precariousness, as reflected upon project-based work and relatively low incomes [see Avdikos and Kalogeresis (2017) for an analysis of incomes of Greek designers in CWS]. So, while for most of digital nomads spatio-temporality is one of the basic ingredients of an individualistic project towards self-realisation and reflexivity, for some others it is the medium of uncertainty that is associated with higher individual risks.

The Need for a Paradigm Shift for CWS Towards the Cooperative Movement

What do these shifts and developments might mean for the future of CWS in Athens, as the digital nomads flows are expected to intensify in the forthcoming years? The answer to this requires us to think about the core function of coworking; is coworking an antidote to precariousness that freelance labour entails or is it a service and an asset tied up in a tourism-led urban growth machine? The recent entanglement of CWS with the tourism sector, as described earlier, poses a series of challenges that directly relate with the continuous distancing of CWS from their initial emergence as spaces that assist freelancers to deal with precarity, while providing them with the conditions and means towards individual and collective well-being, more specifically their shift from production to consumption-oriented practices and services, through processes in which labour-related aspects are increasingly neglected.

Building on the above, a major antithesis is expressed in relevant CWS: while they territorialise flows of digital nomads, at the same time they deterritorialise communities of domestic coworkers, along with established local socio-spatial arrangements through the diffusion of tourists in various segments of the urban environments, including housing, public infrastructure, consumption, etc. The question arising relates to how CWS could align again with the core of the coworking movements' principles, especially concerning addressing the needs of precarious workers. In this vein, re-focusing and re-orientating CWS' operation around issues of labour is of crucial importance towards the empowerment and meeting of their transformative potentials. For us, the notion of 'transformation' is a political one, signifying processes that trigger, accelerate, and enforce social change through the redistribution of power and the emancipation of disadvantaged groups, in our case precarious freelancers and professionals. In this direction, organisational, managerial and governance aspects of CWS are key, along with the need to re-think them in relation with coworkers' well-being and empowerment. In corporate CWS, vague invocations of the notion of community are operating as a means of washing the non-existence of meaningful and transformative arrangements that would allow the coworkers to communicate their common problems, interests and desires in ways that could lead to the emergence of collective identities that could challenge precarity and pave the way for more assertive collectivities of freelancers and professionals. Towards this direction, we argue that valuable lessons can be drowned from the cooperativist movement and social and solidarity economy (SSE) initiatives (Gibson-Graham, 2003). These alternative modes of productive arrangements are based on specific principles and organisational attributes that are of formal character, including regular meetings and assemblies, horizontal governance and decision-making processes, etc.

Such a development could trigger (i) the activation and engagement of locally based freelancers and professionals, (ii) their partaking in decision-making and an upgraded role in both strategic and everyday management of CWS and, thus, the prioritisation of their interests instead of those of other groups of individuals and (iii) the re-articulation of CWS around labour-centred issues and broader societal goals, which could increase their capacities to interact with local settings and to operate as transformative urban actors. More specifically, the co-operative movement is organised around specific principles (International Co-operative Alliance, 2015) which could inform coworking: voluntary and open membership, democratic member control, member economic participation, autonomy and independence, education, training and information, cooperation among cooperatives, concern for community. Among these principles, having coworkers as members who also participate in governance and decision making (according to the one member-one vote rule which is central in both cooperatives and SSE initiatives) could discourage the temporal engagement of coworkers while prioritising their needs. At the same time, member economic participation could create solidarity practices towards supporting members who are struggling with periods of unemployment and the overall implications of a precarious, freelance work status. The autonomous and independent status could prevent the absorption and enclosure of CWS from private, for-profit actors while defining clear rules and limits in synergies and cooperation of CWS with actors from the private sector and the state. Education, training, and information activities could promote collaborative and collective processes of capacity building and skills acquirement for the CWS members, in which individual knowledge and experiences (as well as the labour involved in training activities) could be transformed into common pool resources for the community of coworkers [see also Avdikos and Pettas (2021)]. Finally, the creation of networks among CWS that undertake similar characteristics and attributes, along with the concern for the community (through for example the design and implementation of projects that address local communities' needs and concerns, the enabling of socially innovative ideas and projects, etc.) could lead to the creation of ecosystems that involve both CWS and local actors who, in common, discuss and try to address issues, while exchanging knowledge and good practices that relate on the one hand with the working conditions of the CWS members and on the other hand with local social, spatial and economic conditions.

However, certain limitations could also be associated with the alignment of the coworking movement with the principles of the cooperativist one. CWS which have better chances to be resilient and sustainable, while prioritising the needs of their members over profit are those who will bring together workers who are sharing similar problems and, thus, diversity concerning the range of professions will be limited. Yet, given the increasing deregulation of labour landscapes and a wide range of professions through the two-fold implication of both the digitalisation of labour and the extension of the attributes of 'gig work' and 'platform labour' to conventional professions, such a development could increase the chances of coworkers empowerment through, for example, attempts of unionisation, similar to those emerging in platform and gig work (e.g., Amazon workers, unions of workers in food delivery platforms, etc., Bunders et al., 2022). Workers associations as enablers of CWS cooperatives—enriching their range of interventions and bringing to them a new role that could reconstitute their membership in the emerging landscape of labour deregulation. Additionally, the contribution of the alignment with cooperative principles and organisational modes, can also assist towards preventing grassroots and bottom up CWS from decline due to the predominance of informality concerning the allocation of tasks, duties, and responsibilities, etc. (see Chap. 4 in this book), through the introduction of specific governance arrangements.

Of course, CWS as cooperatives is not something new, as there are a few coworking initiatives that have been developed as such. De Peuter (2017), building on cases of coworking cooperatives in North America and Europe (among others, Cowork Niagara, ECTO, Sackville Commons Cooperative, Nido Coworking and childcare, Workplace Co-operative 115, VillageOffice) argues that such spaces can successfully relate with workers unions and respond to broader political and economic challenges, while bringing out their potential to operate as 'mutual aid accelerators'. The latter, unlike what happens in mainstream, corporate-driven accelerators, could build on experiences from commons initiatives and housing co-ops towards confronting and reversing the 'self-exploitation, IPO-centrism, extractivist venture capital, social inequalities and competitive individualism' trends. In an interview, Cowork Niagara's founder Trevor Twining identifies as the biggest benefit from adopting a cooperative business model the high levels of engagement between coworkers and the space, as the former associate their personal success and development with the progress of the CWS. Building on the previous sections of the chapter, we consider this engagement as crucial for both the creation of meaningful relations and communities and the emergence of CWS as transformative actors in cities.

Conclusion

This chapter focused on the need to re-think CWS' core missions and goals, as recent transformations, such as the pandemic and the digitalisation of labour have augmented remote work arrangements. More and more workers are becoming digital nomads, aiming to combine work and leisure, and usually eager to explore destinations in the Global South and recently in specific destinations in South Europe. Their economic superiority renders them important factors of change in the socio-economic environment of CWS, as CWS usually develop some new additional services to serve the needs of that new strata of coworkers. CWS can act as mediators for digital nomads flows, channelling them and their tourism-led consumption patterns in the recipient city. CWS' mission is then re-augmented towards a more consumption-oriented service provision that can cause a number of disruptions in the rhythm of the internal environment of CWS, exacerbating various socio-spatial inequalities. CWS can be transformed into impersonal service providers for tourist-workers, deterritorialising and displacing the domestic coworkers (and their communities) that usually are in more precarious positions than the digital nomads. From a demand-side perspective, CWS can be driven to become hospitality service providers for a more privileged stratum of coworkers.

In contrast to these new developments in the coworking landscape, we argue that an alignment with the principles of the cooperative movement and the landscape of the Social and Solidarity Economy could reconstitute the CWS with a new role in providing a safety net for a not-so-privileged and precarious workforce, but also in providing the terrain for emancipatory practices of labour solidarity. The outcomes of a cooperativist turn of the coworking movement could reshape the notion of 'community' which has been increasingly employed as an empty signifier and associated with consumption practices within the emergence and spread of 'coworking as a service'. The repositioning of labour in the core of coworking practices, including both the interpretations and portrayals of precarity and attempts to overcome its implications through collective action and claims, as well as in the form of solidarity among CWS members, against individualisation. Furthermore, such a development can maximise the transformative potentials of CWS, through a two-fold focus on production (through the lens of labour) and social reproduction (through practices that revolve around the well-being of coworkers). Finally, the emergence and networking of cooperative-led CWS can transform relevant spaces from facilitators of consumption to urban actors and networks that can operate as hubs for the development and spread of socially innovative practices and democratic values that will contribute to social and economic resilience of the districts and cities in which they are nested.

References

Alexandri, G., & Janoschka, M. (2020). 'Post-pandemic' transnational gentrifications: A critical outlook. *Urban Studies, 57*(15), 3202–3214. https://doi.org/10.1177/0042098020946453

Almeida, J., & Belezas, F. (2022). The rise of half-tourists and their impact on the tourism strategies of peripheral territories. In *Tourism entrepreneurship in Portugal and Spain: Competitive landscapes and innovative business models* (pp. 181–191). Springer International Publishing.

Anderson, B., & McFarlane, C. (2011). Assemblage and geography. *Area, 43*(2), 124–127. https://doi.org/10.1111/j.1475-4762.2011.01004.x

Anderson, B., Kearnes, M., McFarlane, C., & Swanton, D. (2012). Materialism and the politics of assemblage. *Dialogues in Human Geography, 2*(2), 212–215. https://doi.org/10.1177/2043820612449298

Avdikos, V., & Iliopoulou, E. (2019). Community-led Coworking spaces: From co-location to collaboration and collectivization. In R. Gill, A. C. Pratt, & T. E. Virani (Eds.), *Creative hubs in question* (pp. 111–130). Palgrave Macmillan.

Avdikos, V., & Kalogeresis, A. (2017). Socio-economic profile and working conditions of freelancers in co-working spaces and work collectives: Evidence from the design sector in Greece. *Area, 49*(1), 35–42. https://doi.org/10.1111/area.12279

Avdikos, V., & Pettas, D. (2021). The new topologies of collaborative workspace assemblages between the market and the commons. *Geoforum, 121*, 44–52. https://doi.org/10.1016/j.geoforum.2021.02.012

Avdikos, V., Gaitanidis, L., Kalogeresis, A., Kostopoulou, E., Merkel, J., Michailidou, M., Papageorgiou, A., Pettas, D., & Tselepi, N. (2022). *Policy brief 1. Coworking trends in Athens and the impact on the city*. UrbanCoWork Project.

Balampanidis, D., Maloutas, T., Papatzani, E., & Pettas, D. (2019). Informal urban regeneration as a way out of the crisis? Airbnb in Athens and its effects on space and society. *Urban Research & Practice*. https://doi.org/10.1080/17535069.2019.1600009

Bernstein, E., Blunden, H., Brodsky, A., et al. (2020). The implications of working without an office. *Harvard Business Review*, 15 July. Retrieved December 13, 2020, from https://hbr.org/2020/07/the-implications-ofworking- without-an-office?

Bonneau, C., & Aroles, J. (2021). *Digital nomads: A new form of leisure class?* Cambridge University Press.

Borges, I., Brás, S., Machado, A., Leite, S., Costa, E., & Mota, S. (2022). Digital nomads: A growing trend in hospitality in Portugal. In *Advances in tourism, technology and systems: Selected papers from ICOTTS 2021* (Vol. 2, pp. 549–559). Springer Nature.

Bosworth, G., Whalley, J., Füzi, A., & Merell, I. (2021). Rural coworking: "It's becoming contagious". *The Region*. https://doi.org/10.1080/13673882.2021.00001096

Bozzi, N. (2020). # digitalnomads,# solotravellers,# remoteworkers: A cultural critique of the traveling entrepreneur on Instagram. *Social Media + Society*. https://doi.org/10.1177/2056305120926644

Bunders, D. J., Arets, M., Frenken, K., & De Moor, T. (2022). The feasibility of platform cooperatives in the gig economy. *Journal of Co-operative Organization and Management*. https://doi.org/10.1016/j.jcom.2022.100167

Collinge, C. (2006). Flat ontology and the deconstruction of scale: A response to Marston, Jones and Woodward. *Transactions of the Institute of British Geographers, 31*, 244–251. https://doi.org/10.1111/j.1475-5661.2006.00201.x

Dagkouli-Kyriakoglou, M., Tulumello, S., Cocola-Gant, A., Iacovone, C., & Pettas, D. (2022). Digital mediated short-term rentals in the (post-)pandemic city. *Digital Geography and Society*. https://doi.org/10.1016/j.diggeo.2022.100028

De Peuter, G. (2017). *Coworking and co-operation: A union in the making. STIR: The magazine for the new economy*. Retrieved December 13, 2020, from https://culturalworkersorganize.org/wp-content/uploads/2017/10/coworking-and-co-operatives-a-union-in-the-making.pdf.

De Peuter, G., Cohen, N. S., & Saraco, F. (2017). The ambivalence of coworking: On the politics of an emerging work practice. *European Journal of Cultural Studies, 20*(6), 687–706. https://doi.org/10.1177/1367549417732997

Drăgan, M. C., & Camară, G. (2021). Studying tourist congestion as a form of overtourism through Tripadvisor reviews. Case study: Charles bridge, Prague. *Zbornik radova Departmana za geografiju turizam i hotelijerstvo, 51*(1), 39–53. https://doi.org/10.5937/ZbDght2201039D

Ehn, K., Jorge, A., & Marques-Pita, M. (2022). Digital nomads and the COVID-19 pandemic: Narratives about relocation in a time of lockdowns and reduced mobility. *Social Media + Society*. https://doi.org/10.1177/20563051221084958

Eurofound. (2020). *Living, working and COVID-19*. COVID-19 series. Publications Office of the European Union, Luxembourg.

Ferris, T. (2007). The 4-hour workweek. *Fortune Small Business, 17*(4), 84.

Florida, R. (2002). Bohemia and economic geography. *Journal of Economic Geography, 2*(1), 55–71. https://doi.org/10.1093/jeg/2.1.55

Gandini, A., & Cossu, A. (2021). The third wave of coworking: 'Neo-corporate' model versus 'resilient' practice. *European Journal of Cultural Studies, 24*(2), 430–447. https://doi.org/10.1177/1367549419886060

Gaudecker, H. M. V., Holler, R., Janys, L., Siflinger, B. M., & Zimpelmann, C. (2020). *Labour supply in the early stages of the CoViD-19 Pandemic: Empirical evidence on hours, home office, and expectations (No. 13158)*. Institute of Labor Economics (IZA).

Gibson-Graham, J. K. (2003). Enabling ethical economies: Cooperativism and class. *Critical Sociology, 29*(2), 123–161. https://doi.org/10.1163/156916303769155788

Hall, G., Sigala, M., Rentschler, R., & Boyle, S. (2019). *Motivations, mobility and work practices; the conceptual realities of digital nomads*. In information and communication technologies in tourism 2019: Proceedings of the international conference in Nicosia, Cyprus, January 30–February 1, 2019 (pp. 437–449). Springer International Publishing.

International Co-operative Alliance. (2015). *Guidance notes to the co-operative principles*. Retrieved January 15, 2021, from https://ica.coop/sites/default/files/2021-11/ICA%20Guidance%20Notes%20EN.pdf.

Kathimerini. (2019). *Athens established as city break destination*. Retrieved January 15, 2021, from https://www.ekathimerini.com/economy/246983/athens-established-as-city-break-destination/

Law, J. (2004). *After method: Mess in social science research*. Routledge.

Mancinelli, F. (2020). Digital nomads: Freedom, responsibility and the neoliberal order. *Information Technology & Tourism, 22*(3), 417–437. https://doi.org/10.1007/s40558-020-00174-2

Merkel, J. (2015). Coworking in the city. *Ephemera, 15*(2), 121–139.

Merkel, J. (2019). Freelance isn't free.' Co-working as a critical urban practice to cope with informality in creative labour markets. *Urban Studies, 56*(3), 526–547. https://doi.org/10.1177/0042098018782374

Murdoch, J. (1998). The spaces of actor-network theory. *Geoforum, 29*(4), 357–374.

Penney, J., & Dramowicz, K. (2016). Geographical aspects of geo-arbitrage: Work in Canada and live in countries with low cost of living. *IOP Conference Series: Earth and Environmental Science, 34*(1), 012025.

Pettas, D., Avdikos, V., Iliopoulou, E., & Karavasili, I. (2022). "Insurrection is not a spectacle": Experiencing and contesting touristification in Exarcheia, Athens. *Urban Geography, 43*(7), 984–1006. https://doi.org/10.1080/02723638.2021.1888521

Simandan, D. (2017). Competition, contingency, and destabilization in urban assemblages and actor-networks. *Urban Geography, 39*(5), 655–666. https://doi.org/10.1080/02723638.2017.1382307

Spinuzzi, C., Bodrožić, Z., Scaratti, G., & Ivaldi, S. (2019). "Coworking is about community": But what is "community" in coworking? *Journal of Business and Technical Communication, 33*(2), 112–140. https://doi.org/10.1177/1050651918816357

Statista. (2022). *Share of digital nomads worldwide as of March 2022, by income*. Retrieved March 15, 2022, from https://www.statista.com/statistics/1298865/digital-nomads-by-income-level-worldwide/

Thompson, B. Y. (2019). The digital nomad lifestyle:(remote) work/leisure balance, privilege, and constructed community. *International Journal of the Sociology of Leisure, 2*(1–2), 27–42. https://doi.org/10.1007/s41978-018-00030-y

Tselepi, N., Papageorgiou, A., Gaitanidis, L., & Avdikos, V. (2023). Events and their impacts on coworking spaces. *Sustainable Development, Culture, Traditions Journal, 2A*, 2023.

Worlddata.info. (2020). *Tourism in Greece*. Retrieved March 15, 2021, from https://www.worlddata.info/europe/greece/tourism.php.

Part II
Care and Labour in Coworking Practices

Coworking Spaces as Social Infrastructures of Care

Janet Merkel

Introduction

As a shared practice, coworking emerged out of the specific needs of freelance and self-employed workers who have to organise and manage their labour on their own (see Bologna, 2018). Shared workspaces promised to provide a better work infrastructure for independent modes of work and sociality among like-minded workers with similar work-related challenges. In this chapter, I will argue that coworking spaces can also be understood as social infrastructures of care. It aims to explore the analytical potential of the concept of care for understanding experiences and practices of coworking and facilitates a different conceptual lens on coworking. Moreover, it contributes to recent debates in urban geography on care in cities (Alam & Houston, 2020; Greenhough et al., 2023; Power & Williams, 2020).

Over the past decade, a growing body of research has started to interrogate the characteristics of coworking and coworking spaces and whether shared office spaces facilitate social support, a sense of community, new knowledge or even innovation among coworkers (see, for example, Bouncken et al., 2020; Mariotti et al., 2021; Orel et al., 2021). However, the place-based character of these spaces and their social intersections with neighbourhoods and cities have received surprisingly little attention from scholars. Drawing on Klinenberg (2018) arguments, the concept of "social infrastructure" can be used to value these kinds of spaces in their local contexts. Because coworking spaces create "affordances for social connections" (Latham & Layton, 2019, p. 3), they matter as sites for inclusion, belonging and the creation of social relationships. Furthermore, coworking spaces can facilitate the sharing of resources and mutual support and become spaces for care practices in cities. My contribution in this chapter will be twofold. First, it will discuss

J. Merkel (✉)
Institute of Urban and Regional Planning, Technische Universität Berlin, Berlin, Germany
e-mail: janet.merkel@tu-berlin.de

J. Merkel et al. (eds.), *Coworking Spaces*, https://doi.org/10.1007/978-3-031-42268-3_6

coworking spaces as care infrastructures (Alam & Houston, 2020), illuminating the affective, emotional, and embodied dimensions of "coworking". And secondly, it broadens the scope of urban care research (Power & Williams, 2020) and contributes towards this research field.

I begin by discussion those two sets of literature and their central concepts and will explore how they contribute to a better understanding of coworking and coworking spaces and their potential emancipatory role. The chapter ends with a discussion that centers on coworking as a practice able to facilitate meaningful, caring social relationships, supporting empathy and a certain ethics that recognises the interdependencies and builds a strong foundation for social change. I will argue that looking at coworking spaces as an infrastructure of care foregrounds practices and orientations that can lead to emancipatory and transformative politics and new solidarities through the politics of interdependence.

Coworking Spaces as Social Infrastructures

Since Eric Klinenberg's *Palaces for the People* (2018) the notion of social infrastructures has become a prominent concept in urban studies and has inspired new scholarship on how certain public places can facilitate social activities and public life in cities. Klinenberg (2018) argues that public libraries, parks, playgrounds or swimming pools and sidewalks are crucial social infrastructures in cities because they are places where people can assemble, make social connections, and build social relationships which are crucial for the civic life of cities and to help fight inequalities. Because these institutions or places can create "affordances for social connections" (Latham & Layton, 2019, p. 3), they become *social* infrastructures, essential materialities for building community, cooperation, and common purpose (Amin, 2014). It is in this sense that I use the term social infrastructure.

The notion of social infrastructures ties into the "new infrastructural turn" (Amin, 2014, p. 138) that has developed over the last two decades in social and urban theory (Berlant, 2016; Graham & Marvin, 2002, 2022; Larkin, 2013). Thinking with infrastructures questions the role of socio-technical systems and material assemblages in urbanisation processes. It foregrounds the role of the built environment, background materials and structures that allow things to happen and interrogates how they "are implicated in the human experience of the city and in shaping social identities" (Amin, 2014, p. 139). Infrastructures operate in the background, intended to facilitate circulation, living and activity. An infrastructure perspective can help to understand "the affordances that particular spaces or facilities offer for inhabitation and social interaction. It involves looking at the communities and networks of association generated through such spaces" (Latham & Layton, 2019, p. 5). While many public spaces are purposefully created and provided to function as social infrastructures, often the social infrastructural character emerges as a "social surplus" (Latham & Layton, 2019, p. 5) and exemplifies how these institutions can be productive of urban social relations. As social infrastructures these places enable modes of "being together" for people and neighborhood communities that constructs a particular kind of exposure to the larger world (Berlant, 2016). Recent research shows that activities and social connections facilitated through coworking can range from fleeting encounters or temporary affiliations to much stronger bonds, such as friendships (Garrett et al., 2017) or solidarity (Bianchi et al., 2018). Coworkers provide each other with recognition and professional identity (Bacevice & Spreitzer, 2022), the shared use of resources (Avdikos & Pettas, 2021), collective experiences of belonging and community (Garrett et al., 2017) as well as help and mutual support (Gerdenitsch et al., 2016). These research results suggest that coworking spaces matter as sites for inclusion, belonging and the creation of social relationships. However, there is also rising literature that looks towards the darker side of coworking emphasizing the contradictory (de Peuter et al., 2017) and often exclusionary nature of coworking (Lorne, 2020) as well as the affective labour necessary to maintain the social atmosphere and the sense of community that intensifies work for coworkers in coworking spaces (Wright et al., 2022; see also Gatsinos & Höfner in this volume).

So far, little is known under what conditions coworking spaces become social infrastructures and what specific qualities make them *social* infrastructures (Waters-Lynch & Duff, 2021). Research shows that "design, maintenance, distribution, and qualities of what is provided affects how social infrastructures function" (Latham & Layton, 2019, p. 9). Thus, the material and social design of the space plays a crucial role in enabling or constraining social activities.

Coworking Spaces and Care

In coworking research, social relationships in coworking spaces are often considered as transactional, as purposeful interactions that coworkers intentionally seek when using coworking spaces (Bandinelli & Gandini, 2019). Those interactions can emerge from serendipitous encounters while moving around the space but can also come from more engineered encounters through networking events, such as communal breakfast or themed get-togethers. In coworking research less attention has been paid how coworking spaces enable (or constrain) forms of sociality and life to flourish. Drawing on feminist care research and the recent turn to care in geography, I want to argue that coworking spaces can facilitate practices of care in cities or other spatial contexts (Tronto, 1998).

Coworking spaces are social spaces, constituted through myriad relationships inside, outside, and across coworking spaces at different spatial scales. Relational feminist theories of care present a theoretical lens that could provide important insights into coworking and the social relationships as care "begins with a social ontology of connection: foregrounding social relationships of mutuality and trust (rather than dependence)" (Lawson, 2007, p. 3). In this view, care is not seen "as a separate kind of relation, but as endemic to (potentially) all social relations that matter" (ibid). A care ethics also "suggests different ways of theorizing politics" (ibid) as it imagines actors as interdependent and not as autonomous subjects, and thus locates "care's political potential in its relationality" (Bartos, 2020, p. 264).

Recent scholarship in geography has researched how care is practiced in and through specific spaces such as homes, day centers, shelters for the homeless or community food shops as "key sites" of care in the city (Power & Williams, 2020). When we think about how care is practiced in and through particular spaces, we mainly think of institutionalised spaces of care or of households, where woman still carry much of the reproductive labour and the care for others (Federici, 2012; Lynch et al., 2021). However, recent geographic research has extended the notions of care in cities towards various everyday non-institutional care spaces (see, e.g., Bowlby, 2011; Lawson, 2007; Rishbeth & Rogaly, 2018). Coworking spaces can also be considered as sites of care, as places that enable and facilitate mutual support and where "care, as a capacity and a practice, is cultivated, shared and resourced on an egalitarian basis" (Care Collective, 2020, p. 6). First, because coworking spaces are a spatial context for the creation and maintenance of social relationships of independent workers and, second, because they are nurturing reparative work and "minor acts of social repair" (Hall & Smith, 2015, p. 11) in providing guidance, emotional support, and informal help among coworkers and in facilitating acts of "kindness" (Brownlie & Anderson, 2017) in highly individualised work situations. These spaces attend to the needs and well-being of independent workers, and they encourage compassion and mutual help among the coworkers, who receive little attention politically and often find themselves in rather precarious work situations (McRobbie, 2016; Worth & Karaagac, 2021). Nevertheless, care has not been used as an analytic framework in coworking research, despite, for example, much evidence about the emotional and affective labour of coworking spaces' hosts (Merkel, 2019; Waters-Lynch & Duff, 2021) and the immaterial labour of coworkers in creating and maintaining the affective atmospheres and supportive communities within coworking spaces (Bernhardt, 2021; Garrett et al., 2017). Even working from a coworking space is often described as a form of self-care, as having positive benefits for independent workers' well-being because it creates boundaries between work and home (Reissner et al., 2021) and helps them to enact productive work environments (Liegl, 2014).

My thinking about care and infrastructures of care is mainly influenced by recent work in feminist geography whereby care is understood to involve "the proactive interest of one person in the well-being of another and as the articulation of that interest (or affective stance) in practical ways" (Conradson, 2003, p. 508). The Care Collective recently defined as: "Care is our individual and common ability to provide the political, social, material, and emotional conditions that allow the vast majority of people and living creatures on this planet to thrive—along with the planet itself" (Care Collective, 2020, p. 6). Thus, care as understood here does not just mean forms of "hands-on" care such as nursing in hospitals or care homes, teaching in schools or parenting, so all practices that entail "looking after the physical and emotional needs of others" (Care Collective, 2020, p. 5). Rather care is seen much broader as a concern for others, an "other-centredness" (Lynch et al., 2021) and "as a social capacity and activity involving the nurturing of all that is necessary for the welfare and flourishing of life" (Care Collective, 2020, p. 5). In their much-cited definition, Berenice Fisher and Joan Tronto explain:

> On the most general level, we suggest that caring be viewed as a species activity that includes everything that we do to maintain, continue, and repair our world so that we can live in it as well as possible. That world includes our bodies, ourselves, and our environment, all of which we seek to interweave in a complex, life-sustaining web (1990, p. 40).

This definition focuses much on the various practices of care (Puig De La Bellacasa, 2012). But it also contains the idea of a "politics of interdependence" (Care Collective, 2020, p. 5) and a certain ethics of care that develops through caregiving (Tronto, 1993). Many scholars highlight this transformative political potential within care because: "[…] to put care center stage means recognising and embracing our *interdependencies*" (Care Collective, 2020, p. 5; emphasis in original). Thus, this literature is not just highlighting the connectedness but also a mutual dependence of individuals. In this perspective, for example, the care of activists in constructing libraries of things, co-operative alternatives or even housing activists is discussed because in the end "our thriving are everywhere and always contingent on others" (p. 30). Tronto (2015) argues that caring practices foster responsibility, attentiveness, responsiveness, and competence and enable an "ethics of care" that supports empathy and recognizes mutual interdependencies as a foundation for social change.

> Caring requires that one start from the standpoint of the one needing care or attention. It requires that we meet the other morally, adopt that person's, or group's, perspective and look at the world in their terms (Tronto, 1993, p. 19).

From this perspective, care can be seen as embedded within a broader context which resonates particularly well with geographical scholarship. Therefore, coworking spaces as shared relational spaces could constitute a social infrastructure for facilitating care practices among coworkers and for fostering an ethics of care. A care lens might help to direct our attention towards the construction of new forms of relationships, institutions, and action that enhance mutuality and well-being in the context of coworking; explore further the affective, emotional, and embodied dimensions of coworking; extend our understanding of practices of care and gives insights on the form and functions of care that emerge through coworking (i.e., informal care). And it provides useful analytical insights into the practices and politics of interdependence and solidarity and the wider social role that coworking spaces might play in neighbourhoods, cities or beyond (Care Collective, 2020). For this chapter, I use "care" as a lens to highlight how people in coworking spaces engage in subtle acts of care.

My research interest in care is rooted in my empirical investigations on coworking over the last decade. When I talked to coworking hosts they often used the word "mothering" to describe their work practices and to characterize their complex role in the spaces (Merkel, 2015). First mothering seemed an unusual choice to describe their work activities, as it entails self-less giving, nurturing and emotional investment. But then it became clear that a substantial part of their jobs can be described as emotional and affective labour (Hardt, 1999; Hochschild, 2003 [1983]) in attending to the different needs of coworkers, managing their own and coworkers' feelings and in maintaining those welcoming, affective atmospheres (Gregg, 2017; Gregg & Lodato, 2018). Coworkers often describe feelings of "looked out for" or "being cared for" in coworking spaces, as well as being "recognised in their potential" or they talk about "the help they receive to flourish." Most explained that going to work from a coworking space is a form of self-care, as having positive benefits for their own mental wellbeing, because it creates boundaries for the individual worker (Reissner et al., 2021). Yet many mentioned how they struck up friendships with other coworkers and how they become sources of help for each other. Nevertheless, many coworking spaces offer a variety of sport programs, mindful events and workshops or celebrate individual success that can be critically read as "troubling false care" (Bartos, 2021), as supporting the production of neoliberal subjectivities (Bandinelli, 2020) where "caring for *the self* is a pervasive order of biopolitical morality" (Puig De La Bellacasa, 2012, p. 197).

Nevertheless, for this paper, I am interested in situated practices of care—so caring that emerges out of situated needs and not through planned programs, events, or facilitated activities that these spaces create to attract coworkers. I rely on data that I gathered over the last 10 years to understand the development of coworking spaces and the work practices of coworking hosts. As an analytical framework, Tronto's five different phases of care provide a suitable heuristic to analyse different practices of care in a coworking spaces. In their earlier work Fisher and Tronto identified caring as having four analytically separate, but interconnected, phases: *caring about*, *taking care of*, *care-giving* and *care-receiving* (Fisher & Tronto, 1990, p. 40). Recently, Tronto (2015) added a fifth dimension of *caring with* to provide a link between care and democracy and a way to examine the transformative political implications of caring practices. While the first three phases focus mainly on the caregiver's practices, the fourth phase focuses on the care-receiver. Tronto argues that with all the different phases certain ethics and behaviours are practiced that can invigorate a meaningful democracy. And thus provides a way to conceptualize the emancipatory potential of coworking through the practices of situated care.

Practices of Situated Care in Coworking Spaces

In the first phase, *caring about* someone or something asks individuals to recognise and identify a need and to decide whether this need should be met (Fisher & Tronto, 1990, p. 42). Most coworking hosts as well as coworkers care about the space and the people who work from there, and this often extends to the neighbourhood and particular socio-spatial context the coworking space is located in. Caring about does not just mean taking an interest but recognising that care is necessary "to repair and maintain our world" (Fisher & Tronto, 1990, p. 40)—referring here for example to the repair and maintenance of a coworking space as a social and productive work environment for those independent workers. While for coworking hosts the "caring about" orientation lies at the core of their professional hospitality activities and the market service a coworking space offers, it is not for coworkers. Many coworkers develop personal attachment to the space and the other people around—sometimes in a more abstract sense of "caring about the community" but often in more concrete ways (Bernhardt, 2021). Through working side by side, they can become more aware of other coworkers' needs—whether those needs relate to, e.g., quietness for concentration, a friendly chat or help in solving a problem. Many studies have already pointed out that there are often rules emerging around the use of headphones, phones, or video calls and when it is appropriate to talk to someone to respect the need for concentration (Faure et al., 2020). Tronto (2015, p. 7) argues that caring about others results in *attentiveness* for other people's needs.

In the second phase, the *taking care* of "implies the responsibility for initiating and maintaining caring activities" (Fisher & Tronto, 1990, p. 42) and refers to finding the resources to fulfil this responsibility. So, taking care might relate to the willingness of coworkers to engage with others in the space, to understand more about the functioning of the space or to take on more responsibilities for the community. For hosts it can mean to build better knowledge of the different sources of help that should be available for coworkers through the coworking space, building relationships with the coworkers to understand their needs and how to help with different problems. For many hosts it is vital to "sit where coworkers can see them," give the impression of approachability, and engage coworkers in friendly chats. Many hosts said, for example, coworkers wished to bring their dogs into the coworking space which demanded research and negotiation with other coworkers before dogs could be allowed. Also, many coworking spaces engage with their local communities and organise fundraisers, giving and volunteering and need to find local partners. Taking care of someone, or something, produces then *responsibility* (Tronto 2015, p. 7) to oneself, others, places, neighbourhoods, and social communities.

The third phase of *giving care* relates to the actual care work of maintaining and repairing. Here we find a variety of practices such as listening, helping, supporting, cleaning, making coffee/tea, bringing food to share, boundary-making as self-care, organising care (e.g., activities) or informing about more specific help outside the coworking space. Most often it is small acts of situated care between coworkers or coworkers and hosts like bringing a coffee or tea, a friendly word in the morning or an encouraging chat that makes coworkers value the social atmosphere and to engage in caring practices themselves, and thus to take responsibility for the community and the social atmosphere of the space (Bernhardt, 2021). As shown by other scholars this social support and mutual help often extends beyond work-related help into their private life (Garrett et al., 2017; Gerdenitsch et al., 2016; Wright et al., 2022). Also, hosts do not necessarily give the care, very often they connect people and organise appropriate help from inside or from outside the space, relying on their extensive networks. This third phase mainly relates to the work of hosts in coworking spaces who then develop competences in skills and resources in their caring work and can guide others to do the same. However, there are many spaces where coworkers take on much more responsibilities and volunteer to assist with the community management or the organization of specific spaces in the coworking space (such as the library, kitchen, or workshop). Giving care then produces *competence* (Tronto 2015, p. 8) in caring work.

The fourth phase of *receiving care* relates to making sure the care work has been done and has made things better for the care receiver. This phase of care is not only the most difficult because receiving care also means that one must allow others to share in the responsibility of caring for oneself (Fisher & Tronto, 1990, p. 42). In this phase responsive feedback is crucial in establishing a caring relationship. Many coworking spaces have procedures in place to receive feedback from their members such as questionnaires, town hall meetings, or they offer email feedback or a slack channel. Most host engage with coworkers or ask directly if the needs were met and highlight the rather informal ways of care-receiving by checking in over a coffee or the community breakfast. As most caring happens informally also the feedback is given informally to the person who provided it—mainly the host or the other coworkers. However, most hosts emphasise listening carefully to members needs and whether some need was addressed appropriately and satisfactorily. As Tronto (2015, p.8) notes, such a commitment produces *responsiveness*.

The fifth phase of *care with* was developed recently by Tronto (2013): it signifies a practice of communal solidarity that sees care and care responsibility become public concerns (Power & Williams, 2020, p. 5). For example, there are many concerns of freelance workers and self-employed that need to be addresses and should become public concerns across coworking spaces—from disadvantages in social security provisions or the lack thereof, affordable health care to childcare provision in hectic, bulimic work schedules. Working from a shared workspace can help recognising these shared struggles and developing alternatives (Coworking Ideas Project, 2023; Gandini & Cossu, 2021). There are examples where coworking spaces cared with and achieved tangible results for coworkers across several spaces (see Chap. 15 in this book). There is also a rise in initiatives of coworking spaces in rural areas that engage in local development (Gandini & Cossu, 2021). Moreover, man spaces engage in collective efforts of giving more visibility to coworking spaces locally in creating associations or alliances or in making more visible certain professional fields coworkers engage in. And thus, emphasize that coworking spaces too can become a crucial "social infrastructure of interdependent action and solidarity" (Smith, 2020, p. 607).

Coworking Spaces as Social Infrastructures of Care

Coworking spaces as shared relational spaces could provide a social infrastructure for caring practices and relationships and foster an ethics of care among coworkers. This chapter argued that a care lens helps recognise interdependencies among the people involved in coworking practices and fosters that ethic of care, responsibilities, and solidarities. It also helps to see coworking spaces as sites of the everyday practices of situated, informal types of care—and, thus, to think through the socio-political possibilities of coworking that emerge from these everyday practices.

Tronto's (2015) political feminist care ethic offers a way to conceptualise the emergence of such politics as the result of the different care practices and the orientations they foster, as those make people more attentive, responsible, competent, and responsive to others, "heralding different ways of relating" (Lynch et al., 2021, p. 62). The caring practices lay a foundation for a different kind of ethic that underpins relationships in coworking spaces with the recognition of interdependence as sustaining life—a call recently echoed by the Care Collective (2020). Or as Puig De La Bellacasa (2012, p. 20), explains: "With what and, more importantly for this essay, how we relate affects the building of positions and relational ecologies" (Puig De La Bellacasa, 2012, p. 20).

This chapter focuses on caring practices within coworking spaces, and thus on coworking spaces as social infrastructures of care for those who choose to work from shared workspaces. However, coworking spaces can also be imagined as having a wider social infrastructural role in neighbourhoods, cities, or regions. To become these infrastructures, they must create intersections with these different scales and facilitate activities and encounters (e.g., in the neighbourhood context through open workshops, events, canteens and cafés or engaging with local partners or in local activism). Crucial then is the publicness of these spaces and whether they allow these types of interaction and relationships to form. For now, it remains empirically not well understood how the infrastructural role of coworking spaces can be extend to the wider public.

A care-lens highlights the affective, emotional, and embodied dimensions of "coworking" as lived everyday practices that that have been rarely addressed in coworking research (see Chap. 9 in this book). Furthermore, it points to various practices of giving and taking care by coworking host and coworkers that these spaces can facilitate. Moreover, it broadens the scope of recent urban care research (Power & Williams, 2020) beyond spaces of interpersonal care (e.g., homes) and institutional care (e.g., elderly homes, schools, kindergarten) and discusses shared workspaces as a space of care. In line with recent geographical scholarship, it extends our understanding of practices we usually discuss as care, as it highlights more situated caring practices that emerge out of situational needs of coworkers.

If we think coworking spaces as caring institutions, then we also must acknowledge the specificities of the context and the politics of care in coworking spaces. This chapter intentionally focused on the potentials of a care lens, but shared workspaces are of course spaces where independent workers' pay to be "cared for". This characteristic provokes question about inequalities and forms of social exclusion in access to care practices for independent workers in coworking spaces (Coworking Ideas Project, 2023; Lorne, 2020; Wright et al., 2022). Following Fisher and Tronto (1990, p. 40), on could ask whose worlds are being maintained, continued, and repaired, and whose worlds are not? Several scholars already pointed out that coworking might reproduce existing inequalities and reinforce boundaries of inclusion and exclusion (Burchell et al., 2021; Leung & Cossu, 2019; Rodríguez-Modroño, 2021. With a rising number of independent workers and remote workers, future research needs to address these inequalities more thoroughly. Thus, a *critical care approach* (Bartos, 2021) to scrutinise whether caring practices and the creation of caring relations inside of coworking spaces facilitates transformative strategies for social emancipation and economic alternatives (Gibson-Graham, 2006). Although "caring as a service" in coworking spaces comes with a clearly structured relationship between caregiver (host or community manager) and the care receiver (coworker), there is empirical evidence where this relationship became multidirectional and extended into friendship-making and gift-exchange between the hosts and coworkers, or where coworkers regularly take on responsibilities for the space and each other (Bernhardt, 2021). Many coworking spaces provide crucial social, material, and emotional support for freelance and self-employed workers to thrive through their care practices. The different types of relationships facilitated through coworking add to the "sources of intimacy, care and support that people use to get through the vicissitudes of life, both major and minor" (Bowlby, 2011, p. 618). In fostering and shaping social relationships and sociality of independent and remote workers, coworking spaces can be sites where caring orientations and relationships are produced and reproduced. Care provides a valuable lens into alternative narratives, emancipation politics and the wider social role that coworking spaces might play in neighbourhoods, cities, or regions.

References

Alam, A., & Houston, D. (2020). Rethinking care as alternate infrastructure. *Cities, 100*, 102662. https://doi.org/10.1016/j.cities.2020.102662

Amin, A. (2014). Lively infrastructure. *Theory, Culture & Society, 31*(7–8), 137–161. https://doi.org/10.1177/0263276414548490

Avdikos, V., & Pettas, D. (2021). The new topologies of collaborative workspace assemblages between the market and the commons. *Geoforum, 121*, 44–52. https://doi.org/10.1016/j.geoforum.2021.02.012

Bacevice, P. A., & Spreitzer, G. M. (2022). 'It's like, instant respect': Coworking spaces as identity anchoring environments in the new economy. *New Technology, Work and Employment*. https://doi.org/10.1111/ntwe.12254

Bandinelli, C. (2020). The production of subjectivity in neoliberal culture industries: The case of coworking spaces. *International Journal of Cultural Studies, 23*(1), 3–19. https://doi.org/10.1177/1367877919878449

Bandinelli, C., & Gandini, A. (2019). Hubs vs networks in the creative economy: Towards a 'collaborative individualism'. In R. Gill, A. Pratt, & T. Virani (Eds.), *Creative hubs in question. Dynamics of virtual work* (pp. 89–110). Palgrave Macmillan.

Bartos, A. E. (2020). Relational spaces and relational care: Campus sexual violence, intimate geopolitics and topological polis. *Area, 52*(2), 261–268. https://doi.org/10.1111/area.12449

Bartos, A. E. (2021). Troubling false care: Towards a more revolutionary 'care revolution' in the university. *ACME: An International Journal for Critical Geographies, 20*(3), 312–321.

Berlant, L. (2016). The commons: Infrastructures for troubling times*. *Environment and Planning D: Society and Space, 34*(3), 393–419. https://doi.org/10.1177/0263775816645989

Bernhardt, A. (2021). *Coworking-Atmosphären. Zum Zusammenspiel von kuratierten Räumen und der Sicht der Coworkenden als raumhandelnde Subjekte*. Springer VS. https://doi.org/10.1007/978-3-658-35888-4

Bianchi, F., Casnici, N., & Squazzoni, F. (2018). Solidarity as a byproduct of professional collaboration: Social support and trust in a coworking space. *Social Networks, 54*, 61–72. https://doi.org/10.1016/j.socnet.2017.12.002
Bologna, S. (2018). *The rise of the European self-employed workforce*. Mimesis International.
Bouncken, R., Ratzmann, M., Barwinski, R., & Kraus, S. (2020). Coworking spaces: Empowerment for entrepreneurship and innovation in the digital and sharing economy. *Journal of Business Research, 114*, 102–110. https://doi.org/10.1016/j.jbusres.2020.03.033
Bowlby, S. (2011). Friendship, co-presence and care: Neglected spaces. *Social & Cultural Geography, 12*(6), 605–622. https://doi.org/10.1080/14649365.2011.601264
Brownlie, J., & Anderson, S. (2017). Thinking sociologically about kindness. Puncturing the Blasé in the ordinary City. *Sociology, 51*(6), 1222–1238.
Burchell, B., Reuschke, D., & Zhang, M. (2021). Spatial and temporal segmenting of urban workplaces: The gendering of multi-locational working. *Urban Studies, 58*(11), 2207–2232. https://doi.org/10.1177/0042098020903248
Care Collective. (2020). *The care manifesto: The politics of interdependence*. Verso.
Conradson, D. (2003). Spaces of care in the city: The place of a community drop-in centre. *Social & Cultural Geography, 4*(4), 507–525. https://doi.org/10.1080/1464936032000137939
Coworking Ideas Project. (2023). *Making coworking more inclusive, diverse, equitable and accessible*. Retrieved February 23, 2023, from https://coworkingidea.org/
de Peuter, G., Cohen, N. S., & Saraco, F. (2017). The ambivalence of coworking: On the politics of an emerging work practice. *European Journal of Cultural Studies, 20*(6), 687–706. https://doi.org/10.1177/1367549417732997
Faure, S., Aroles, J., & De Vaujany, F.-X. (2020). At the heart of new work practices: A paradoxical approach to silence in a coworking space. *Ephemera: Theory and Politics in Organization, 20*(4), 307–322.
Federici, S. (2012). *Caliban und die Hexe. Frauen, der Körper und die ursprüngliche Akkumulation*. Mandelbaum Verlag.
Fisher, B., & Tronto, J. (1990). Toward a feminist theory of caring. In E. Abel & M. Nelson (Eds.), *Circles of care: Work and identity in women's lives* (pp. 35–62). SUNY Press.
Gandini, A., & Cossu, A. (2021). The third wave of coworking: 'Neo-corporate' model versus 'resilient' practice. *European Journal of Cultural Studies, 24*(2), 430–447. https://doi.org/10.1177/1367549419886060
Garrett, L. E., Spreitzer, G. M., & Bacevice, P. A. (2017). Co-constructing a sense of community at work: The emergence of community in coworking spaces. *Organization Studies, 38*(6), 821–842. https://doi.org/10.1177/0170840616685354
Gerdenitsch, C., Scheel, T. E., Andorfer, J., & Korunka, C. (2016). Coworking spaces: A source of social support for independent professionals. *Frontiers in Psychology, 7*(581). https://doi.org/10.3389/fpsyg.2016.00581
Gibson-Graham, J. K. (2006). *A postcapitalist politics*. University of Minnesota Press.
Graham, S., & Marvin, S. (2002). *Splintering urbanism: Networked infrastructures, technological mobilities and the urban condition*. Routledge.
Graham, S., & Marvin, S. (2022). Splintering urbanism at 20 and the "infrastructural turn". *Journal of Urban Technology, 29*(1), 169–175. https://doi.org/10.1080/10630732.2021.2005934
Greenhough, B., Davies, G., & Bowlby, S. (2023). Why 'cultures of care'? *Social & Cultural Geography, 24*(1), 1–10. https://doi.org/10.1080/14649365.2022.2105938
Gregg, M. (2017). *From careers to atmospheres. CAMEo Cuts #3*. CAMEo Research Institute for Cultural and Media Economies, University of Leicester.
Gregg, M., & Lodato, T. (2018). Managing community: Coworking, hospitality and the future of work. In B. Röttger-Rössler & J. Slaby (Eds.), *Affect in relation. Families, places, technologies* (pp. 175–196). Routledge.
Hall, T., & Smith, R. J. (2015). Care and repair and the politics of urban kindness. *Sociology, 49*(1), 3–18. https://doi.org/10.1177/0038038514546662
Hardt, M. (1999). Affective labor. *boundary 2, 26*(2), 89–100.

Hochschild, A. R. (2003 [1983]). *The managed heart: Commercialization of human feeling*. University of California Press.
Klinenberg, E. (2018). *Palaces for the people. How social infrastructure can help fight inequality, polarization, and the decline of civic life*. Crown.
Larkin, B. (2013). The politics and poetics of infrastructure. *Annual Review of Anthropology, 42*(1), 327–343.
Latham, A., & Layton, J. (2019). Social infrastructure and the public life of cities: Studying urban sociality and public spaces. *Geography Compass, 13*(7), e12444. https://doi.org/10.1111/gec3.12444
Lawson, V. (2007). Geographies of care and responsibility. *Annals of the Association of American Geographers, 97*(1), 1–11. https://doi.org/10.1111/j.1467-8306.2007.00520.x
Leung, W.-F., & Cossu, A. (2019). Digital entrepreneurship in Taiwan and Thailand: Embracing precarity as a personal response to political and economic change. *International Journal of Cultural Studies, 22*(2), 264–280. https://doi.org/10.1177/13678779188212
Liegl, M. (2014). Nomadicity and the care of place—On the aesthetic and affective organization of space in freelance creative work. *Computer Supported Cooperative Work (CSCW), 23*(2), 163–183. https://doi.org/10.1007/s10606-014-9198-x
Lorne, C. (2020). The limits to openness: Co-working, design and social innovation in the neoliberal city. *Environment and Planning A: Economy and Space, 52*(45), 747–765. https://doi.org/10.1177/0308518x19876941
Lynch, K., Kalaitzake, M., & Crean, M. (2021). Care and affective relations: Social justice and sociology. *The Sociological Review, 69*(1), 53–71. https://doi.org/10.1177/0038026120952744
Mariotti, I., Di Vita, S., & Akhavan, M. (Eds.). (2021). *New workplaces—Location patterns, urban effects and development trajectories: A worldwide investigation*. Springer.
McRobbie, A. (2016). *Be creative: Making a living in the new culture industries*. Wiley.
Merkel, J. (2015). Coworking in the city. *Ephemera: Theory & Politics in Organization, 15*(1), 121–139.
Merkel, J. (2019). Curating strangers. In R. Gill, A. Pratt, & T. Virani (Eds.), *Creative hubs in question. Dynamics of virtual work* (pp. 51–68). Palgrave Macmillan.
Orel, M., Dvouletý, O., & Ratten, V. (Eds.). (2021). *The flexible workplace. Coworking and other modern workplace transformations*. Springer. https://doi.org/10.1007/978-3-030-62167-4
Power, E. R., & Williams, M. J. (2020). Cities of care: A platform for urban geographical care research. *Geography Compass, 14*(1), e12474. https://doi.org/10.1111/gec3.12474
Puig De La Bellacasa, M. (2012). 'Nothing comes without its world': Thinking with care. *The Sociological Review, 60*(2), 197–216. https://doi.org/10.1111/j.1467-954X.2012.02070.x
Reissner, S. C., Izak, M., & Hislop, D. (2021). Configurations of boundary management practices among knowledge workers. *Work, Employment and Society, 35*(2), 296–315. https://doi.org/10.1177/0950017020968375
Rishbeth, C., & Rogaly, B. (2018). Sitting outside: Conviviality, self-care and the design of benches in urban public space. *Transactions of the Institute of British Geographers, 43*(2), 284–298. https://doi.org/10.1111/tran.12212
Rodríguez-Modroño, P. (2021). Non-standard work in unconventional workspaces: Self-employed women in home-based businesses and coworking spaces. *Urban Studies, 58*(11), 2258–2275. https://doi.org/10.1177/0042098021007406
Smith, T. S. (2020). 'Stand back and watch us': Post-capitalist practices in the maker movement. *Environment and Planning A: Economy and Space, 52*(3), 593–610. https://doi.org/10.1177/0308518X19882731
Tronto, J. C. (1993). *Moral boundaries: A political argument for an ethic of care*. Routledge.
Tronto, J. C. (1998). An ethic of care. *Generations: Journal of the American Society on Aging, 22*(3), 15–20.
Tronto, J. C. (2013). *Caring democracy: Markets, equality, and justice*. NYU Press.
Tronto, J. C. (2015). *Who cares? How to reshape a democratic politics*. Cornell University Press.

Waters-Lynch, J., & Duff, C. (2021). Coworking's cooperation paradox: On the role of stigmergic curation. In M. Orel, O. Dvouletý, & V. Ratten (Eds.), *The flexible workplace: Coworking and other modern workplace transformations* (pp. 133–156). Springer International Publishing. https://doi.org/10.1007/978-3-030-62167-4_8

Worth, N., & Karaagac, E. A. (2021). Accounting for absences and ambiguities in the freelancing labour relation. *Tijdschrift voor economische en sociale geografie*. https://doi.org/10.1111/tesg.12491

Wright, A., Marsh, D., & Wibberley, G. (2022). Favours within 'the tribe': Social support in coworking spaces. *New Technology, Work and Employment, 37*(1), 59–78. https://doi.org/10.1111/ntwe.12214

Coworking Spaces in Aotearoa New Zealand: Embodied Geographies of Care in Pandemic Times

Joke J. Methorst

Introduction

> Fundamentally, in the new version of it and the new iteration of it, it's like, well, this is what our people have been doing for 1000s of years. So, living in tribes and living in clans, living in villages. And over the years, we've kind of moved, we've kind of migrated away from each other. And then what happens is, there's no collaboration, the connectivity is a bit more fractured. So, bringing people back into an environment that you could meet cool new people, likeminded people, share ideas (Rebecca, Māori female, mid-40s, 29 April 2021).

Although coworking spaces (CWSs) vary in size, scope, and geographical location, they often evoke a desire for human connectedness, for embodying space, and for geographies of care. This connectivity is epitomised in the operator's assertion that social connectedness and belonging have always been a basic need. Building on recognition of this desire, I argue that alongside the dominant coworking values of collaboration, innovation, networking, and knowledge exchange in a world driven by technologies, the disruptive nature of CWSs can also create places of care that can nurture embodied geographies for greater connectivity and inclusivity (Fig. 1).

Considering CWSs as places where embodied geographies of care can be practiced and produced, not only reveals the practices of care taking, but also provides an alternative narrative to the 'celebratory framework that surrounds the representations of proliferating coworking spaces' (Gandini, 2015, p. 193). The disruptive nature that characterises much of CWSs' contribution in the reconfiguration of work, also has the potential to transform and reshape normative notions of work organisation and can provide an alternative insight into the embodied caring practices (Lawson, 2007).

J. J. Methorst (✉)
University of Waikato, Hamilton, New Zealand
e-mail: jm365@students.waikato.ac.nz

J. Merkel et al. (eds.), *Coworking Spaces*, https://doi.org/10.1007/978-3-031-42268-3_7

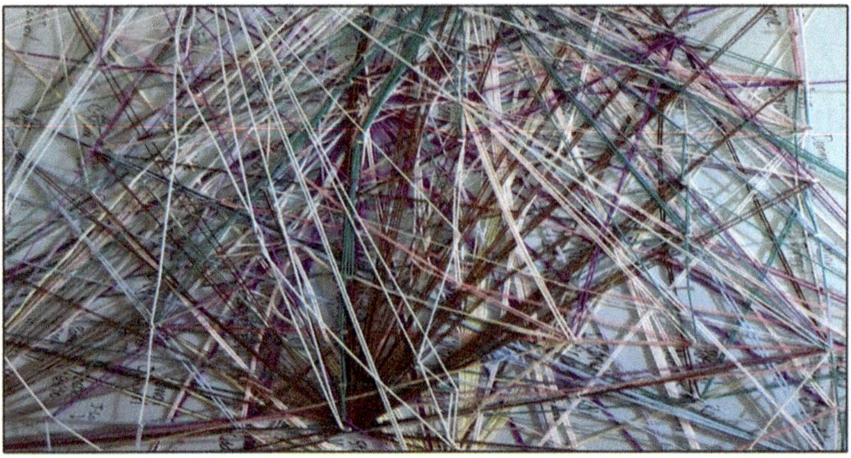

Fig. 1 Human connectedness exemplified by pieces of string on a whiteboard in a CWS. (Image by author)

In what follows, I introduce an embodied geographies of care approach and demonstrate how this perspective can broaden the understanding of CWSs as alternative spaces of care. After discussing the methodology, the empirical findings of the chapter then illustrate how CWSs in Aotearoa New Zealand articulate through different embodied caring practices by notions of inclusivity, safety, and care. Some of the notions of inclusivity are articulated and filtered through principles that emerge within Te Ao Māori (the Māori worldview) such as manaakitanga (hospitality, kindness, generosity), kaitiakitanga (guardianship) and koha (donation), that are part of many people's (Māori, Pacifica, Pākehā, and others) everyday geographies in Aotearoa New Zealand. These expressions of care that are manifested in the CWSs, can be viewed as part of a broader care ethics that surrounds the reshaping of normative notions around the organisation of work(place). These changes are deeply embedded in the neoliberal capitalist society and dominate current CWS practices. As such, the chapter demonstrates the transformative potentials of CWSs as places of care and the emancipatory potentials of coworking.

Geographies of Care: The Transformative Potential of CWSs

The mobility of work that characterises CWSs, is often associated with globalising forces of technological development, neoliberal individualism, liquid modernity and network capital (Elliott & Urry, 2010). Underpinning this development lies a hegemonic discourse on digital technology (Richardson, 2017), which marginalises and devalues principles of care, affect and emotion in a dominating capitalist neoliberal society (Tronto, 2017). Coworking is illustrative of how digital technologies, along with their attendant political rationalities and economic configurations, influence multiple geographies of work and workplace, and shape the everyday socio-spatial experiences of people (Richardson, 2021). Not only have the work/leisure and private/professional life boundaries become increasingly blurred, but the disruptive technologies that have led to a mobilisation of work, workplace, and workers, also challenged the normative understanding of being 'at' work (Richardson, 2018). Disruptive innovation is a term that is used to describe how innovative products can transform traditional dominating industries by offering new alternatives (Guttentag, 2015), of which for example the radical growth of Airbnb and Uber illustrate how conventional industries have been challenged. The disruptive nature of CWSs can thus be understood in terms of how the proliferation of CWSs challenges established norms and normative understanding of conventional office space.

In contrast to accounts focused on the disruptive nature of CWSs in blurring organisational boundaries (Howell, 2022), reconfiguring traditional workspaces, and constituting fluid organisational arrangements (Brinks, 2022), there has been very little research from feminist and embodied geography perspectives. Instead, I argue that the disruptive nature of CWSs also has the potential to produce places of care in work environments—that are still characterised by disembodiment, neutrality and rationality, devaluing bodies, subjectivity, and emotion (Eisler & Eisler, 2008). An embodied geographies of care approach can shed light on how 'working space is felt through disruption of the boundaries of the intimate sphere through digital technologies' (Richardson, 2018, p. 257). To understand the geometries of power at play, feminist geographers focus on the 'event of place' (Massey, 2005, p. 130). How care is structured, practiced, narrated and felt in CWSs is understood as co-produced, or in Massey's words, within a 'constellation of processes' (Massey, 2005, p. 141). Following Massey (2005), CWSs are conceived as continuously evolving places that both shape and are shaped by understandings and notions of identity, making people feel in and/or out of place.

I argue that, especially in a neoliberal individualised society, CWSs have the transformative potential to create more embodied geographies of care and inclusivity. An embodied geography framework gives space to people, practices and places that are largely marginalised by dominating Western thought. As such, the disruptive nature that characterises CWSs can be explored from another side; one that focuses on principles of care, affect and emotion, to which I turn next.

Embodied Geographies of Care

An embodied geography perspective is particularly helpful in exploring the power geometries of CWSs. Work on an ethos of care in geography is flourishing (Adams-Hutcheson & Johnston, 2019), which is based on care as a universal need (Tronto, 2017). Several feminist scholars have advanced a focus on practices of care and highlighted the need for care ethics to be incorporated into the epistemological, ontological and methodological dimensions of scholarship (Tronto, 2017). When, however, practices of care continue to be situated in a capitalist neoliberal society where human wellbeing is based on neoliberal logics of efficiency, standardisation and competition (Harvey, 2007), the normative notion of ethics of care continues to be devalued (Tronto, 2017). Feminist geographer Massey (2005) explains normative ethics of care as a Russian-doll geography of affect, where the territorial nature of geographies of care is still rooted in hegemonic thinking, as the need and responsibility for ethics of care is based on spatial proximity. The further away, the less important and less strong geographies of affect are. This is also noticed by Lawson (2007), who argues that rather than framing care as a privatised and territorialised construct governed by neoliberal principles, feminist ethics of care 'foreground[s] social relationships of mutuality and trust (rather than dependence) [...] understands all social relations as contextual, partial, attentive, responsive, and responsible' (p. 3). With feminist theories of care, normative ideas about the individualised self in the neoliberal economy can be disrupted (McDowell, 2004), as it reveals the relationality of spaces of care, where practices of care traverse any space, blurs boundaries and where everyone shares the same responsibility. Puig de la Bellacasa, therefore, calls for researchers to attend to how care in places circulates 'not necessarily morally or intentionally, [but] in an embodied way, or simply embedded in the world, environments, infrastructures that have been marked by that care' (2017, p. 121). This chapter responds to Puig de la Bellacasa's call to identifying caring places by extending the notions of care in places through an examination of care practices within CWSs.

Infrastructures of care 'constitute dynamic structures and practices that organise social and material lives' (Alam & Houston, 2020), and can be found in purposefully designed caring spaces for people who have a specific need for care (Power & Williams, 2020), but also in non-institutional everyday spaces. A study by Lawrence and Maitlis (2012) for instance, found that organisations are potential spaces of compassion, where embodied geographies of care are enacted through the everyday practices within those organisations. Such an approach is particularly important in relation to a post-pandemic world dominated by digital technologies and oppressive global capitalism (William, 2020), where caring practices become the antidote in response to crisis (Bryson, 2021). McDowell (2004) sees the ethics of care as a countermove against the increasingly individualised neoliberal capitalist world, where the interdependent practices of solidarity and mutuality can be encouraged.

Practices of care can be applied onto different existing research areas. Within the domain of geography, Power and Williams (2020) examine and identify care in spaces and materialities in cities. By looking at objects and bodies, they examine how these materialities shape the nature of care and how care is performed in particular spaces constituting places of care. Their analysis thus expands existing research on urban care that regards cities as an assemblage of interpersonal caring practices, by 'arguing that care has the potential to be present in all urban spaces and places and that care has transformative potential in these settings' (pp. 6–7). The practices of care within any given place shape the socio-spatial dimensions of that place and people embodying that space and vice versa. Care, in this regard, is 'inextricably relational' (Conradson, 2003, p. 508). Looking at CWSs through an embodied geographies of care perspective, allows for an understanding of the transformative social potential of CWSs. As I argue, CWSs can be spaces of care where operators can create more inclusive environments and hospitable spaces and where they move away from the capitalist mindset underpinning much of the commodification of CWSs at present time. By framing CWSs as spaces of care, the insights from this chapter illustrate the diversifying practices of CWS operators in Aotearoa New Zealand and illuminate the transformative potential of CWSs, where they become an example of a more diverse economy and, ultimately 'could draw on the pleasures of friendliness, trust, and companionable connection' (Gibson-Graham, 2008, p. 619).

The Contribution of This Chapter

The contributions of this chapter are threefold: First, this research opens up avenues for considering CWSs as spaces of care by extending the notion of embodied geographies of care perspective onto the domain of shared workspaces. As such, the chapter contributes to existing research by exploring everyday practices of care in CWSs as extended spaces of care. Second, it looks at CWS operators as potential agents of change who can support empathy by curating a space that fosters mutual support and solidarity by for example providing childcare onsite. The 'koha economy' articulates a view that people working from CWSs can draw on resources that are not necessarily monetised. By moving away from institutionalised spaces, CWS operators challenge standardised norms in the workplace that articulate with ethics of care, creating more inclusive and emancipatory environments. Third, extending the particular focus of research on European CWSs (Berbegal-Mirabent, 2021), this chapter expands CWS scholarship by providing unique empirical insights from CWSs in Aotearoa New Zealand where practices of care articulate with notions of inclusivity, safety and care. Within the unique setting of a postcolonial nation-state, some of these notions of inclusivity are articulated and filtered through principles that emerge within Te Ao Māori (the Māori worldview) such as manaakitanga, kaitiakitanga and koha. As such, the empirical data provide exciting insights that not only extend the understanding of practices of care in CWSs, but also how, in an increasingly digitalised, post-pandemic world, the affective and emotional qualities of CWSs emerge through embodiment of space.

CWSs in Aotearoa New Zealand

Aotearoa New Zealand, a geographically isolated island nation-state in the Pacific, has a population of over 5.1 million inhabitants and a high urbanisation rate of 86.79% in 2021 (Stats NZ, 2021). The country's ethnicity is diverse: 70.2% of the people identified themselves with European ethnicity, 16.5% Māori, 15.1% Asian and 8.1% with Pacific ethnicity (Stats NZ, 2018). The official languages are English, Te Reo Māori and New Zealand Sign Language.

The first CWS in Aotearoa New Zealand, BizDojo,[1] appeared in the country's largest city Auckland in 2009 after which the company soon expanded to multiple spaces in Auckland, Wellington and Christchurch. CWSs in Aotearoa New Zealand have since then proliferated, as from merely three CWSs taking up 1.400 m^2 of shared workspace in 2011 to 76 CWSs covering 79.203 m^2 in the three main cities of Auckland, Christchurch and Wellington in 2022 (Bayleys, 2022).

The average monthly price for a permanent desk for a member is $509.91 NZD whereas a day pass averages $32.80 NZD (based on available information of all listed CWSs). There is a wide variation in individual rates and perks offered with memberships. For instance, 'Digital Nomad', a boutique urban CWS in Wellington CBD asks $755.00 NZD for a dedicated desk per month, whereas CWS 'Silverfox' in Whangārei charges $240.00 NZD, situated in a more provincial urban location and at the lower end of the market.

Methodology

The data for this chapter were obtained from ethnographic observations, visual data and semi-structured interviews with 11 CWS operators in Aotearoa New Zealand. The rich account of visual data allowed for analysing the look and feel of the CWS as well the outside of the building and location (Rose, 2016). The interviews took place on-site between March and May 2021, which afforded me the opportunity to observe, sit and work alongside other coworkers after the interview had finished. The selected CWSs all varied in size, scope, and number of members (from nine to 160 members) and were located in Auckland, Dunedin, Tauranga and Christchurch. The gender of members varied, from 30 up to 60% being identified as women, as well as the type of industries the coworkers worked in, though creative industries and technology dominated. The gender of the operators varied too, six CWSs had male operators, four female and one CWS was managed by a heterosexual couple. The CWS operators identified mostly with Pākehā (white settlers/European), one identified as Māori and one as Pacific. As one operator wanted to stay anonymous and given the relatively small amount of CWSs in Aotearoa New Zealand, I have decided to not share the names or exact locations. The interviews gathered in-depth data that were used to explore expressions of embodied geographies of care in the existing socio-spatial environments of CWSs.

[1] BizDojo was a New Zealand owned start-up, and after going into liquidation bought by global company IWG (NZ Herald, 2018, February 5).

CWSs as Spaces of Care

CWSs in Aotearoa New Zealand have many affinities with CWSs internationally, which are the result of the commodification and globalisation of CWSs and their emergence as a new trend in work geographies. What the findings in this chapter demonstrate however, is that *some* CWSs in Aotearoa New Zealand are also articulated through notions of inclusivity, safety, and care. Building on the recognition that the disruptive nature of CWSs has transformative potential, CWSs form part of the broader social context of social relations and meaningful dialogues that are attributes of ethics of care. The next sections highlight the three different ways that embodied geographies of care emerge in CWSs.

Creating More Inclusive Environments

CWSs, as any workplace, can be seen as work environments in which gender identities are socially constructed (McDowell, 2011). Building on the notion that traditional workspaces and organisations are gendered sites, 'wherein male advantage is built into job design, distribution of power, rewards and resources, physical design of the workspace, and rules for work-related behaviour' (Sargent et al., 2020, p. 7), the ways in which CWSs are curated are often underpinned by a masculinist rationality. In addition, curated norms and values that shape embodied experiences in CWS environments, play an important part of how operators, by turning to geographies of care, can emphasise feminine values of care. After all, the seemingly open, purportedly public workplaces of CWSs are paradoxical, as CWSs often only cater for a limited range of people and identities, co-producing what are desired bodies and practices in space (Ishitsuka, 2020). As the following quote illustrates, working from a CWS is a lifestyle-based decision that people can make who are not constrained by, for example, caring duties, and who have the 'desired' profession:

> Yeah, a lot of the casuals [are women], we've got a few mums at work here as well, you know, with three days a week work out of here. And it's just getting away from the home environment. Which, you know, is good for a period of time, but it's nice to, I mean, a number of the people, especially the programming people, they can work from home (Paul, NZ Pākehā male, mid-40s, 27 April 2021).

It is perhaps not surprising that women are largely underrepresented as permanent fulltime members and form the largest proportion of 'casuals'. Despite an increasing proportion of (white) women working from western CWSs (Foertsch, 2019), the ways in which bodies are organised in the playful workspace setting of CWSs, often reproduces the young masculine body as the desired body in space. It appears that together with the institutional, classed, and gendered character of the industries coworkers are employed in, there are certain gendered constraints caused by the ability of working remotely from a CWS. The following quote illustrates how women, mostly, are restricted by caring duties:

> Yeah, it's all those little things, it's our goal to make it more accessible for women. There's a woman upstairs who's a civil engineer. And she was saying how, when she was doing her course, she'd have her baby and her course started at eight in the morning. And day-cares don't open until like 830 or 9am. When she was going to day-care, she was always sick. And she couldn't take any time off because you just fall so behind and we're just talking about how it's [refers to conventional workspaces] not accessible (Alex, NZ Pākehā female, late 20s, 29 April 2021).

The ways in which women tend to have more caring duties than men, leads to restrictions of choice where to work from (e.g., home versus a CWS), creating further gendered barriers to work from a CWS. In contrast to conventional workspaces that are designed to be childfree, CWS owners increasingly seem to challenge these standardised norms that articulate with ethics of care. Many CWS operators acknowledge the difficulty for coworkers in dealing with complicated caring duties and work-life balance. Rather than reconstituting these roles, CWS operators move away from institutionalised spaces by welcoming children into the CWS. The following two operators explain:

> Sometimes, parents say they have to bring their kids in and we're okay with that ... And I guess for us it's also creating that environment. The reality is you're gonna have a child that has to stay home from work or school. And I guess we have created spaces for them to be able to do that too (Angela, Pacific female, late 20s, 29 March 2021).

> So, we have actually two single dads upstairs and they have children ... And they come in all the time for the full day. Just hanging out on the couches and play games and that sort of thing ... yeah, people bring their children all the time ... So it's interesting to say, and I guess you get that with marginalised communities [referring to those with caring duties], they want to help people like their own ... So that's our goal as well, making sure that it's accessible for everyone here and that people don't feel like excluded from the space. Because that's when businesses in cities succeed right, when everyone, not just rich, straight and all [succeeds] (Alex, NZ Pākehā female, late 20s, 29 April 2021).

These two operators exemplify that by changing dominant normative structures at the workplace around caring duties and children at workplaces, CWS operators can de-emphasise some dominant masculine values that are not the inherent nature of any CWS. The meaning of working from a CWS then, moves beyond the wider masculine constructed image of a male, young, independent ICT worker (Sargent et al., 2020). The ways in which some CWS operators are reshaping existing gendered relations by creating more inclusive environments, can make a change in the oftentimes marginalised position of women in workplaces and is one of the ways in which practices of care are articulated.

Hospitable Spaces

As CWSs provide a blended workspace with characteristics and amiability of café-like environments, they blur the boundaries of work and private life manifested through specific practices of hospitality. Welcoming people into the space, introducing coworkers to each other, inductions and refill coffee and tea pots, are all performances of hospitality, predominantly by appointed female community managers, reconstituting stereotypical gender roles. Normally CWSs are arranged in a way where coworkers have to abide by certain kinds of rules based on common understandings of expected norms and behaviour. The vision, values, etiquettes and guidelines of a CWS are often explained, and signed upon, in a membership agreement. The following quote demonstrates how working from a CWS means being in a place that is defined by rules and order, to ensure a smooth-running business:

> So, this [pointing to the agreement] is basically and overly dramatically referred to as the Magna Carta. Because it's four levels deep, it has vision values. But then you know, everybody has vision values, right? But then the last two parts are the most crucial. That is the etiquette. Which I guess is sort of situational driven guidelines. So, what's our view, and it talks about what is gossip, what is not gossip etc. And then there's the guidance, so that makes the rules. But we based it on the manaakitanga [being hospitable] principles, with years and years of etiquette based on trust (Cliff, NZ Pākehā male, mid-40s, 7 April 2021).

This standardised practice reinforces coworking to be individualistic and transactional, based on masculine notions of productivity, formality, and achievement, reproducing neoliberal capitalist values. This is indicative of a particular approach in which ethics of care is expressed in a contractual agreement, creating conformity by following the rules set out. The practices of care become symbolised in protocols, which bears a masculine imprint. What the following CWS operator demonstrates however, is an embodied geography of care approach where care ethics articulates manaakitanga. She explains how the manaakitanga values she has been brought up with, have influenced the way she is managing the CWS:

> So kaitiakitanga [guardianship] is looking after our environment, respecting and loving it. And so, everyone that walked in, and there was never a sign up saying anything. There's no, you know, rules other than don't pretend ... There's no magic in the fact that if you make people proud of the space. Selling the atmosphere and change is their game … And we kind of exhale with that, because I think with it being a new concept, it was almost like we wanted to, really, I guess, I'm big on, you know, for when it comes to hospitality or manaakitanga. It's kind of what we do. You know, we're brought up that way of offering people a cup of tea, you know, coffees, you know, whatever it is that you did, the first thing you do when it's what we teach our rangatahi [youth] as well. It's like, greet (Rebecca, Māori female, mid-40s, 29 April 2021).

The operator illustrates how she creates a hospitable environment by practices that embody and articulate with different forms and principles of care, of which specific instances of hospitality are expressed through ideas emergent with Te Ao Māori. The values of being hospitable and inviting people in, is for the operator the normative understanding of what working in a shared workspace entails and embedded in her everyday geography. Ethics of care, in Te Ao Māori, 'is embodied in the cultural thinking and practices associated with whakapapa, mana and manaaki […]. For Māori, recognition of mana and the extension of aroha, is tika – it is right and ethical behaviour – and does not depend on being returned or acknowledged' (Hoskins, 2012, p. 91). Rather than ethics of care documented in a signed agreement, this CWS provides an alternative expression of hospitality, which is based on reciprocity and caring for others, undocumented and without expecting anything in return.

Manaakitanga 'ensures principled practice as the concept orientates us away from locating "deficits" within the person and towards addressing deficits in societal systems' (Hodgetts et al., 2021, p. 1983). By turning to care ethics, this CWS not only disrupts the masculine-based nature that surrounds the organisation of work by attending to Indigenous voices, but also foregrounds social relationships of trust and mutuality, turning to the relationality of spaces of care.

'Koha Economy'

CWSs predominantly cater to knowledge workers in the gig-economy, offering different kinds of membership fees that gives access to the space and the community, based on neoliberal capitalism. These memberships undoubtedly create geographies of inclusion and exclusion, accessible to the privileged with sufficient financial resources, while marginalising those who cannot afford membership. By making membership affordable in other ways, however, CWSs can disrupt dominating neoliberal capitalist forces, indicative of alternative practices of care.

CWS operators and entrepreneurs have, as any other business owner, been directly affected by the global COVID-19 pandemic. After the first lockdown in 2020 however, many CWS operators provided unique ways to keep supporting their coworkers, of whom many struggled due to the precarious nature of their jobs. The following CWS operator in a popular lifestyle area, explains how he, as most of the interviewees, suspended membership fees for existing members who had financial problems and could not afford to come back into the space:

> I'd say over half of our members didn't come back. They didn't have the funds to come back. But what was interesting about that was, we said to a few of them, or actually all of them: if money is the problem, then don't make it the problem. Just come back. Don't pay rent. Until you can. For us, it's more important to have people in the space then not (Richard, NZ Pākehā male, mid-40s, 7 April 2021).

This example demonstrates how the CWS operator creates an embodied geography of care environment, where he puts the financial incentive aside, in which he shares responsibility for the wellbeing of the coworker. The choice to demonetise during times of financial hardship and uncertainty, is another example of how notions of care can be found in extended places of CWSs. Another CWS operator has taken this one step further, as the CWS now operates on the concept of the 'koha economy' (by donation), as she wanted to include people into the space who were not able to pay for membership:

> So previously, we had set memberships, permanents, casuals, flexies, and then hot-deskers. But what we've done now, this time, is that we based our memberships on koha economy … So how the koha economy works is that we give them the opportunity to be like: Hey, this is what we can offer to be able to be a member of this space and there's other things that work around your memberships is that we ask them, that they contribute to monthly events … It could be anything, it'd be something from team building or sharing knowledge or something that he's learned recently, and then another part of the koha economy is that

Fig. 2 Whakataukī at CWS in NZ. (Image by author)

> you contribute to filling up this chocolate jar or, bringing or paying for the milk this week (Angela, Pacific female, late 20s, 29 March 2021).

The idea of a koha economy articulates a view that people using the CWS not only have different financial resources, but also other kinds of resources that are not monetised. Coworkers contribute to the CWS itself, rather than being valued on their income that is used to pay for access to the coworking community. In addition, rather than categorising people into abstract units by using different coworker typologies such as member, resident, permanent or casual, this CWS moves away from terminology to disestablish obvious distinctions in terms of the value of different coworkers, which determines and affects not only the identities that make up the CWS, but the entire constitution of the space. By providing alternative practices of care, this CWS demonstrates how the disruptive nature of CWSs has the potential for a more embodied geography, as this whakataukī (Māori proverb) (Fig. 2) on the wall of one of the CWSs visited implies:

By at least in part, turning towards the koha economy and making membership affordable in other ways, there is an indication that practices and principles of care can be found in CWSs as extended spaces of care, and provides an alternative voice to the dominating neoliberal capitalist forces at play.

Conclusion

In this chapter, I have argued that CWSs, as inviting, open—yet private—workplaces have the potential to act as places of care, as they can play a significant role in the lives of coworkers. By exploring embodied geographies of care that can be found in *some* CWSs in Aotearoa New Zealand, this chapter has highlighted an underappreciated potential of CWSs that evolve around the practices of care.

CWSs predominantly illustrate how digital technologies transform the very nature of work and geographies of workplaces. As CWSs play a central role in how new modes of work unfold by epitomising the reconfiguration of workspace, the insights offer a wider perspective on the values of coworking in post-pandemic times, which can help to inform future CWS research. Embodied geographies of care are demonstrated by changing norms and behaviours that shape the embodied experiences of coworkers, by for instance creating more inclusive environments for those who are usually marginalised in work environments. The ways in which some CWS operators are reshaping existing gendered relations can make a change in the oftentimes marginalised position of women in workplaces and is one of the ways in which practices of care are articulated. The insights also contribute to a diverse understanding of the transformative potential of CWSs, as operators acting as agents of change, invest in caring relationships based on receptivity and reciprocity, coproducing spaces of care. Rather than continuing to marginalise feminist voices in much of CWS scholarship, a focus on principles of care, affect and emotion shows another side of CWSs and the ways in which their disruptive nature can lead to more embodied, caring spaces.

Thinking through an embodied geographies of care perspective, is witnessing transformative potentialities of, in this case CWSs in Aotearoa New Zealand, but can be applied to many more research streams and emerging phenomena. An embodied and emotional geography approach can illuminate, especially in a post-pandemic world, practices of care in the unexpected immaterial and material spaces, ultimately leading to more embodied geographies. By integrating an ethics of care perspective, it can create an understanding that 'care has the potential to be present in all urban spaces and places and that care has transformative potential in these settings' (Power & Williams, 2020, pp. 6–7).

References

Adams-Hutcheson, G., & Johnston, L. (2019). Flourishing in fragile academic work spaces and learning environments: Feminist geographies of care and mentoring. *Gender, Place & Culture, 26*(4), 451–467. https://doi.org/10.1080/0966369X.2019.1596885

Alam, A., & Houston, D. (2020). Rethinking care as alternate infrastructure. *Cities, 100*, 102662. https://doi.org/10.1016/j.cities.2020.102662

Bayleys (2022). *Flexible office options provide freedom of choice for businesses.* Available at https://www.bayleys.co.nz/insightsanddata/commercial

Berbegal-Mirabent, J. (2021). What do we know about co-working spaces? Trends and challenges ahead. *Sustainability, 13*(3), 1416. https://doi.org/10.3390/su13031416

Brinks, V. (2022). Fluid objects? An attempt to conceptualise the global rise of "coworking spaces". *Area.* https://doi.org/10.1111/area.12778

Bryson, S. A. (2021). An ethic of care? Academic administration and pandemic policy. *Qualitative Social Work, 20*(1–2), 632–638. https://doi.org/10.1177/1473325020973386

Conradson, D. (2003). Spaces of care in the city: The place of a community drop-in centre. *Social & Cultural Geography, 4*(4), 507–525. https://doi.org/10.1080/1464936032000137939

Eisler, R., & Eisler, R. T. (2008). *The real wealth of nations: Creating a caring economics.* Berrett-Koehler Publishers.

Elliott, A., & Urry, J. (2010). *Mobile lives: Self, excess and nature.* Routledge. https://doi.org/10.4324/9780203887042

Foertsch, C. (2019). 2019 State of Coworking: Over 2 Million Coworking Space Members Expected. *Deskmag*. Available at https://www.deskmag.com/en/coworking-news/2019-state-of-coworking-spaces-2-million-members-growth-crisis-market-report-survey-study

Gandini, A. (2015). The rise of coworking spaces: A literature review. *Ephemera: Theory and Politics in Organizations, 15*(1), 193–205.

Gibson-Graham, J. K. (2008). Diverse economies: performative practices for 'other worlds'. *Progress in Human Geography, 32*(5), 613–632. https://doi.org/10.1177/0309132508090821

Guttentag, D. (2015). Airbnb: disruptive innovation and the rise of an informal tourism accommodation sector. *Current issues in Tourism, 18*(12), 1192–1217. https://doi.org/10.1080/13683500.2013.827159

Harvey, D. (2007). *A brief history of neoliberalism*. Oxford University Press.

Hodgetts, D., Rua, M., Groot, S., Hopner, V., Drew, N., King, P., & Blake, D. (2021). Relational ethics meets principled practice in community research engagements to understand and address homelessness. *Journal of Community Psychology, 50*(4), 1980–1992. https://doi.org/10.1002/jcop.22586

Hoskins, T. K. (2012). A fine risk: Ethics in Kaupapa Maori politics. *New Zealand Journal of Educational Studies, 47*(2), 85–99.

Howell, T. (2022). Coworking spaces: An overview and research agenda. *Research Policy, 51*(2), 104447. https://doi.org/10.1016/j.respol.2021.104447

Ishitsuka, A. M. (2020). Visual encounters in global Shanghai. On the desirability of bodies in a coworking space. *China Perspectives, 2020*(3), 39–46. https://doi.org/10.4000/chinaperspectives.10416

Lawrence, T. B., & Maitlis, S. (2012). Care and possibility: Enacting an ethic of care through narrative practice. *Academy of Management Review, 37*(4), 641–663. https://doi.org/10.5465/amr.2010.0466

Lawson, V. (2007). Geographies of care and responsibility. *Annals of the Association of American Geographers, 97*(1), 1–11. https://doi.org/10.1111/j.1467-8306.2007.00520.x

Massey, D. (2005). *For space*. Sage.

McDowell, L. (2004). Work, workfare, work/life balance and an ethic of care. *Progress in Human Geography, 28*(2), 145–163. https://doi.org/10.1191/0309132504ph478oa

McDowell, L. (2011). *Capital culture: Gender at work in the city*. John Wiley & Sons.

Power, E. R., & Williams, M. J. (2020). Cities of care: A platform for urban geographical care research. *Geography Compass, 14*(1), e12474. https://doi.org/10.1111/gec3.12474

Puig de La Bellacasa, M. (2017). *Matters of care: Speculative ethics in more than human worlds* (Vol. 41). University of Minnesota Press.

Richardson, L. (2017). Sharing as a postwork style: Digital work and the co-working office. *Cambridge Journal of Regions, Economy and Society, 10*(2), 297–310. https://doi.org/10.1093/cjres/rsx002

Richardson, L. (2018). Feminist geographies of digital work. *Progress in Human Geography, 42*(2), 244–263. https://doi.org/10.1177/0309132516677177

Richardson, L. (2021). Coordinating office space: Digital technologies and the platformization of work. *Environment and Planning D: Society and Space, 39*(2), 347–365. https://doi.org/10.1177/0263775820959677

Rose, G. (2016). *Visual methodologies: An introduction to researching with visual materials*. Sage.

Sargent, A. C., Yavorsky, J. E., & Sandoval, R. G. (2020). Organizational logic in coworking spaces: Inequality regimes in the new economy. *Gender & Society, 35*(1), 5–31. https://doi.org/10.1177/0891243220974691

Stats NZ (2018). *Census 2018 New Zealand.* Available at https://www.stats.govt.nz/tools/2018-census-place-summaries/new-zealand

Stats NZ (2021). *Subnational population estimates (urban rural)*. Available at https://nzdotstat.stats.govt.nz/wbos/Index.aspx?_ga=2.15140594.596012437.1693754122-1634502971.1693754122

Tronto, J. (2017). There is an alternative: Homines curans and the limits of neoliberalism. *International Journal of Care and Caring, 1*(1), 27–43. https://doi.org/10.1332/239788217X14866281687583

William, I. (2020). Global capitalism post-pandemic. *Race & class, 62*(2), 3–13. https://doi.org/10.1177/0306396820951999

You Know That You've Succeeded in Your Role When Your Work Renders You Invisible: The Invisible Work of Community Management

Antigoni Papageorgiou and Martha Michailidou

Introduction

This chapter provides an analysis of the different forms of work conducted by community managers in four coworking spaces in Athens and Berlin. Based on in-depth interviews with community managers and participant observation in these spaces, the chapter examines the role of community managers in the fluid, flexible, and informal organisational contexts of such spaces. The chapter takes its cue from previous studies which call for a more empirically grounded, critical, study on the phenomenon of coworking (de Peuter et al., 2017; Virani & Gill, 2019). It attempts to respond to this call, drawing particularly upon research on the gendered character of relational labour.

Our study revolves around two interconnected set of research questions: first, we examine what the daily tasks and routine of community managers are, as job descriptions are scarce in a sector that reflects the informal ethos of Cultural Creative Industries (CCIs). Second, we turn to the relational practice (Fletcher, 1998) which is part of the job and constitutes the 'community building' process, crucial for the maintenance of the space. We argue that the labour devoted to community building is invisible labour (Daniels, 1987). As a process, it builds on emotional and caring skills which are not formally recognised as skills and offers services which lie at the intersection of typical and atypical work within coworking spaces. Based on these

A. Papageorgiou (✉)
Department of Economic and Regional Development, Panteion University of Social and Political Sciences, Athens, Greece
e-mail: ant.papageorgiou@panteion.gr

M. Michailidou
Department of Communications, Media and Culture, Panteion University of Social and Political Sciences, Athens, Greece
e-mail: marthami@panteion.gr

J. Merkel et al. (eds.), *Coworking Spaces*, https://doi.org/10.1007/978-3-031-42268-3_8

characteristics, the chapter argues that feminist analyses of feminized work are most pertinent for analysing this labour.

The chapter is divided into three sections. The first section offers a brief outline of the empirical studies on coworking and especially of analyses of coworking spaces as spaces that give visibility to a currently 'workplaceless' and, thus, invisible labour force. This section situates our analysis of coworking in relation to issues of labour precarity and the emergence of gig work. The second section focuses on feminist arguments for the redefinition of work; its aim is to draw from analyses of women's unpaid reproductive labour and critical organisation studies in order to shed light on the relational practices which are part of community management. These practices, having historically been carried out by women and, thus, feminised, are not acknowledged as 'real work'. The last section concentrates on the empirical findings of our study. It outlines the relational tasks that community managers have to undertake and discusses the gendered nature that renders those forms of work invisible. This condition of invisibility, resulting from both the gendered nature of these tasks that lie within the sphere of social reproduction and from the lack of sustained organisational structures and processes, restrains community managers from understanding themselves as *workers* with specific job tasks, who are entitled to certain rights, and could organize themselves collectively. Emotion management is more likely to be experienced as part of the self rather than part of the job as the lack of immediate control and direct supervision creates an illusion of autonomy, independency, and ownership – conforming to long-standing work patterns of low status service work.

Coworking Spaces: Giving Visibility to the Precarious, 'Gig' Workers

Scholars have addressed the phenomenon of flexible workplace set-ups, including coworking spaces, from different perspectives. Operating under various hybrid formats, coworking spaces primarily provide shared office space for those who are 'workplaceless' and, thus, not visible as part of the official labour force. Empirical studies of the phenomenon (Avdikos & Kalogeresis, 2016; Merkel, 2019b; Reuschke et al., 2021) argued that coworking made an emerging segment of these professionals, the freelancers, visible. The exponential growth of digitally mediated and enabled work, intensified since the 2008 crisis, has led researchers to advocate for a wide range of issues: from clarifying the definitional boundaries of different employment statuses to raising insurance, health, and safety concerns for those involved in those forms of work (Huws et al., 2018). The increase of subcontracting normalized the creation of flexible workforce which is on demand and ready to be supplied when it is needed. But while freelance work is gradually penetrating all sectors of the economy beyond the creative ones (Mould et al., 2014), it remains relatively invisible in economic geography and employment policies as it is insufficiently measured, understood, and thus inadequately regulated and protected. As Worth and Karaagac (2021, p.96) note, 'Unlike the visibility of minimum wage workers in retail and food service, the precarity of many freelancers is obscured (measured in end products like articles rather than an hourly rate)'.

Studies about gig work (Dunn, 2020; Prassl, 2018) and platform economies (Bosma, 2022) have given freelance work visibility, emphasizing the ways in which labour is being digitally mediated, (re)distributed, and detached from physical workplaces, carried out in temporary offices and ephemeral workplaces. Some of these studies shed light into the intersection of entrepreneurship and home-based work (Webster & Zhang, 2020). Recent empirical studies argue that coworking spaces function as enclaves of precarious work (Pacchi & Mariotti, 2021), where the gig economy is organized around a chain of value creating activities that maintain a pool of human and non-human resources to be used in project-based employment. Following the argument of dePeuter et al. (de Peuter et al., 2017), those scholars shed light into what they termed the 'ambivalence of coworking' – a practice that emphasizes sharing and mutuality but operates within the market economy. In those contexts, coworkers are, in fact, treated as customers who seek particular services – both material and immaterial – and should be kept at least satisfied, if not animated.

Remote, digital, and freelance workers seek shelter in coworking spaces from the isolation and the monotony of working from home and the loneliness which is part of the condition of creative labour (Gill, 2007). What seems to differentiate these spaces from the usual office space facilities is their peculiar 'collaborative atmosphere' (Merkel, 2015). Repeatedly put under academic scrutiny, this collaborative spirit and the related sense of belonging to a community of like-minded individuals seem to be among the top reasons professionals join such spaces. This collaborative atmosphere seems to be quite significant in maintaining their wellbeing, in acquiring new skills, and in expanding their personal and professional networks (Papageorgiou et al., 2022). Research has thus examined how and under what circumstances these shared arrangements could provide a basis for collaboration, knowledge sharing, and even collective organisation and bargaining for a diverse range of professionals that work remotely and independently.

A Request for Visibility: Addressing the Invisible Work of Community Managers

Research on coworking points out that the increase of coworking places has also led to their gradual differentiation into corporate and community-led ones (Avdikos & Iliopoulou, 2019) and argues that the latter nurture community and collaboration as opposed to the profit-driven, top-down entrepreneurial ventures. Claims that coworking could potentially lead towards the fabrication of emerging forms of sociality are pervasive in the literature but insufficiently substantiated. Additionally, the emphasis on serendipity and coworking spaces as 'serendipity accelerators' obscures the fact that coworking spaces are managed communities. In response to the overemphasis on serendipity, recent coworking studies focus on the role of community managers in facilitating interactions, as spatial proximity within a coworking space does not in itself guarantee productive forms of sociality. Merkel (2019a) introduces the concept of social curation, which includes the staging of a wide range of social interactions initiating meaningful connections among coworkers. These staged interactions require planning and strategic thinking from the community managers' side and are seen to foster a collective sense of belonging. Such practices have even been described by managers as 'mothering' (Merkel, 2015, 2019a), a loaded term that connotes selfless giving with no expectation of getting something in return.

Drawing upon empirical studies (Gregg & Lodato, 2018; Merkel, 2019a; Moriset, 2014) that accentuate coworking community managers' practices as crucial for the maintenance of coworking spaces as social infrastructures, we, thus, explore the everyday – visible and invisible – tasks community managers undertake in four coworking spaces. To do that, we make use of feminist readings of labour that expand our understanding as to what counts as work and why some forms of work are rendered invisible. We, thus, explore the relational practices that community managers undertake and employ the concept of visible/invisible labour to capture the gendered character of these practices.

Through a close exploration of the everyday working life of community managers from four coworking spaces, we argue that the transient nature of coworking (Merkel, 2019a) with its loose organisational structure (Blagoev et al., 2019) prevents us from fully capturing the role and tasks of community managers. These tasks are part of a laborious process that involves the warmth and caring necessary for the construction and maintenance of interpersonal relations. Since in their daily tasks community managers lack immediate supervision and control, they see their work as an extension of their personality – this, in return, offers them a sense of independence and autonomy. This independence, which can occasionally also be found in low status service work (McDowell, 2009), is, however, illusory as they have to run the space according to specific guidelines. Moreover, servicing others can take up many forms of work that, when acknowledged, are regarded as particularly suitable for women, mobilising an 'idle capacity' (Webster & Zhang, 2020) where workers, and especially female ones, are using their already existing physical assets to accomplish the job. This relational work, centred in fostering and maintaining social relations, is given fragmented visibility – as we explain in the next section.

Tracing the Invisible (Women's) Work

The concept of invisible labour emerged as a way to discuss the overly narrow preconceptions of what is understood and acknowledged as work (Daniels, 1987). While 'visible labour' might sound redundant as a term – someone could argue that all forms of work are visible – it reveals the gendered roots of common shortcomings around work which does not conform to a very specific pattern: the type of work that is considered hard and takes place at a physical workplace and in formal organisations with fixed structures, processes, and hierarchies and in the public sphere. The type of work that requires training, experience, mental effort, and capacity is valued as highly skilled, and thus, it is assigned with status and prestige. In industrialized societies, work is something for which workers are getting paid for, with the identity of workers, though, being historically very narrow and gendered. Drawing largely upon the Fordist paradigm of work and men's experiences at work, which is deeply rooted in the social construction of the public/private dichotomy and the breadwinner representation of the worker, work has been linked to full-time employment that has advanced nets of social security and organized representation.

Historically, men were occupying the public sphere with their endeavours being visible, acknowledged, and appreciated, while women were undertaking housework that was condemned to remain hidden behind the closed doors of private households. Housekeeping along with childcare and care activities was lacking validation and became partly visible when middle-class women refused to execute them in view of entering the world of paid employment in the service economy outside the household. Care work has gradually been delegated as lower status work and remains entirely undervalued, financially uncompensated, poorly protected, and precarious (Ferrant et al., 2014). In the moral order of a masculinized world of work, care work is seen having altruistic motivations and as fundamentally linked to women's 'natural' abilities for doing the 'dirty work' (Duffy, 2007).

The feminist movement argued for the recognition and thus the remuneration of housework. In Power of the Women and the Subversion of the Community, Dalla Costa and James (1972, p. 5) observe that:

> With the advent of capitalism, the socialisation of production was organised with the factory as its centre. Those who worked in the new productive centre, the factory, received a wage. Those who were excluded did not. Women, children, and the old lost the relative power that derived from the family's dependence on their labour, which was seen to be social and necessary.

One of the greatest examples of feminist filmmaking, 'Jean Dillman' (1975) by Chantal Ackerman, depicts the standardized routine of housework and the sense of alienation and distance from it. Mostly filmed inside the domestic space through a real-time representation of everyday tasks, the film gives visibility to the 'female sphere' and its burdens. Echoing the work of Sylvia Federici which thoroughly criticises the way advanced capitalist economies refused to recognise 'reproductive' tasks as labour (Cox & Federici, 1975; Federici, 2006), feminist thought argues for the redefinition of labour so that women's unpaid reproductive labour is recognised and, crucially, acknowledged as central for the reproduction of labour power and 'a terrain of accumulation and therefore a terrain of exploitation' (Federici, 2006).

Arlie Hochschild's work (1979, p. 551) draws our attention to the way 'emotion can be and often is subject to acts of management'. In her work, 'Emotion Work,

Feeling Rules, and Social Structure' (1979, p. 561), Hochschild explicates emotion work as:

> the act of trying to change in degree or quality an emotion or feeling. To 'work on' an emotion or feeling is, for our purposes, the same as 'to manage' an emotion or to do 'deep acting.' Note that 'emotion work' refers to the effort, the act of trying – and not to the outcome, which may or may not be successful.

Hochschild analyses the labour carried out by low status service workers such as secretaries and airline stewardesses, whose jobs require high degree of emotion and display management. The gendered nature of such practices prevents us from seeing them as 'real work' and thus renders them invisible, making them a 'natural' extension of character.

As organisational researchers observe, society assigns relational activities to women because stereotypically those are considered 'feminine'. According to Joan Acker, 'the gender segregation of work, including divisions between paid and unpaid work, is partly created in organizational processes' (Acker, 1990, p. 140). Relational theory emerged aiming to give visibility and legitimacy to a wide range of practices that were seen as inherent to the feminine and private spheres. Fletcher (1998) analyses relational practice into four categories by creating a new language of relational competence: preserving, mutual empowering, achieving, and creating team. Practices that have traditionally been considered feminine, condemned to remain and not coded as work, are now given visibility and are included into the definition of work and competence.

Methodology and Data Collection

The chapter is based on research carried out in four coworking spaces in Athens and Berlin. Our case studies varied in terms of size, internal organisation, and overall profile (Table 1). In-depth interviews with community managers, founders, and

Table 1 Cases and their characteristics

Name	Size	Community managers	Hosts	Interviews in total	City
Coworking Space 1	Small (10–40 coworkers)	1 – male	n/a	17	Athens
Coworking Space 2	Large (30–60 coworkers)	1 – female	1 – female 1 – male	17	Athens
Coworking Space 3	Very small (10–20 coworkers)	None	None	11	Berlin
Coworking Space 4	Large (30–100 coworkers)	1 – female 1 – male	1 – female 2 – males	11	Berlin

users were carried out in all these spaces as well as participant observation of everyday practices and interactions within the spaces. In two larger coworking spaces under investigation, we also interviewed hosts, situated lower in the organisational hierarchy, who provide support to the community manager and undertake the day-to-day menial tasks of community management and hosting. Coworking space 3, the smaller coworking space among our case studies, operated as a 'self-maintained space', allocating facility management tasks to its users. As we will explore in the next section, the aforementioned space was solely offering shared office space and was explicitly not interested in nurturing social relationships within the space. The relations taking place there were mostly transactional, as none of the founders was willing to undertake a series of relational practices that could cultivate a sense of belonging. Therefore, in this space we didn't identify any collaborative aspects. This finding reinforces our argument that socialisation, collaboration, and even knowledge sharing do not occur serendipitously and organically but as the result of a laborious process.

Community Management: Recruitment and Job Responsibilities

The community managers we interviewed had rich working biographies involving different projects in many countries even though they were all in their first career steps. They nevertheless found it difficult to articulate the exact tasks of their jobs and could be reluctant to discuss their working and employment conditions. None of our interviewees was part of a formal representative organisation or union. Instead, informal networks were crucial to finding new projects. Exploring the working experiences of those who work as community managers in coworking spaces, we came across different levels of informality characterising hiring processes, on-the-job activities, and precarity of employment conditions.

Coworking spaces seem to be organized around a continuum of formal and informal processes (Blagoev et al., 2019), reflecting long-standing patterns of informality and precarity that prevail in the wider CCIs and strengthening pre-existing gendered inequalities (Conor et al., 2015) but also the increasing professionalisation of the coworking sector. Some of our interviewees were informally recruited through word of mouth and indirect network contacts conforming to the pattern of 'network sociality' prevalent in the creative sector (Wittel, 2001). Reflecting on the way he entered the world of coworking, the community manager of Coworking Space 1 notes: 'I met the founder of the space at a terrace party, and this was when I was told that she was looking for someone to take over her responsibilities'. Similarly, another interviewee told us that he was hired during an event organized by Coworking Space 4, where he got to meet one of the founders of the space.

Others were hired under short-term contracts, traineeships, and internships, and thus, the work was low paid and precarious. A female community manager started first as an intern through an EU-funded internship programme and then was hired to undertake the community manager role. This finding, indeed, suggests that there is a reserve pool of labour across Europe whose mobility is facilitated through various programmes that ease the transition to the labour market:

> [...] it was an EU funded education programme, so I basically applied for this programme as an intern project manager, and then it went super well. I don't know, I was so surprised, because it was my first introduction to the job career, like my first time ever I would say. [...] Then, I stayed around, and finally [I was] located there as a community manager. (Female Community Manager of Coworking Space 4)

The female community manager of Coworking Space 2 went through a very long hiring process which included a series of interviews with the founders and the community management team. Despite this aspect of formality and standardisation of the hiring procedure, the research participant had difficulties in articulating the very nature of her occupation. When we asked her for a research interview for our study, she came back to us with a request:

> Ok, I agree to do the interview to talk about my job, and my work at [Coworking Space 2] and hopefully I can get something in return. Maybe you could help me understand really what I'm doing here. I've been trying to get it since the very first day here. (Female, Community Manager, Coworking Space 2)

This reluctance revealed the absence of a formal job description, clearly assigned responsibilities, tasks, as well as line management. The community manager of the larger coworking space in Berlin narrated to us that in her current role as community manager she has no job description nor has ever received any form of training:

> I haven't received any formal training or induction. I'm doing it the best way I can. (Female, Community Manager, Coworking Space 4)

A certain degree of informality and fluidity can be also found in the job responsibilities of community managers. When asked to describe their everyday tasks, community managers tended to focus on the visible, tangible, aspects of their work:

> I never thought of it, because when I got hired my responsibilities were fluid, a bit vague. But as I'm working here, I'm gradually understanding my role: I have to open and set up the space, prepare the space for meetings, to reply to leads and prospect clients, to answer emails [...]– to prepare the social media posts, to check google analytics, to make banners, to manage payments and keep track of memberships, to answer phone calls, to be an active member of the hub network we participate as [Coworking Space 1], to do the newsletter of the space, to seek funding and mobility opportunities, to seek job/project opportunities and share them internally. (Male, Community Manager, Coworking Space 1)

The need for multi-tasking, project management skills, and client service tasks reveals a role that coalesces features of work in the tourism and hospitality sectors as well as the creative work. An interviewee compared his working reality to the one he had when he was working as a bartender:

> I've learned a lot from my previous job as barman. [....] It feels like I'm working in a pub on St. Patrick's Day. (Male, Community Manager, Coworking Space 3)

Participants described an intense working pace with highly fragmented tasks and responsibilities. This, in fact, restrains individuals from having a coherent understanding of their role and a concrete professional identity:

> I actually do a little bit of that, of this... I'm doing almost everything here: from inducting new members to the rules of the space to making coffee and managing the online presence of the space. (Male, Host, Coworking Space 2)

The apparent lack of formal responsibilities creates conditions of obscurity as salaries are more likely to be products of individual negotiations, and any discussion around work benefits is entirely absent. The inherent informality, and the subsequent ephemerality that short-term work contracts bring, creates a vacuum in which community managers must interpret and carry out their own tasks on the basis of personal intuition and individualised responsibility, blurring the boundaries of work and non-work.

As observed, conflicts occur when new people get hired to undertake community manager roles. Indicative was the case of a young female worker who refused to undertake cleaning and maintenance responsibilities, thereby implicitly challenging the gendered expectations which come as part of the job. While her female employer never asked her male colleague to perform such tasks, she was expecting from her to respond as woman and do what is considered as 'women's work'. Similar expectations were also expressed from coworkers when a female host asked a resident to leave his office clean:

> One day I got into a huge argument with a resident. He was leaving all of his trash behind. I reported it but nothing happened, the guy kept on doing that. He secretly liked the fact that I was chasing him and telling him to follow the rules. (Female, Host, Coworking Space 4)

When realised, the gendered burden that women in these occupations face is managed through essentialist solutions, in the hope that the replacement of women by men in such positions would minimise traditionally gendered expectations and defeminise the job. As the female founder of Coworking Space 1 explained to us, she deliberately looked to hire a male for the position of community manager:

> [name of the Community manager] got hired because I got fed up having women community managers. Looking back, we had a huge problem with coworkers asking her to do their dirty work, like wash their cups and dishes. We were like two women managing the community, I desperately wanted to hire a man. (Female Founder, Coworking Space 1).

Relational Practices

When pressed for more details regarding their working realities, community managers narrated to us individualised tactics they personally employ to execute their relational tasks with the right attitude:

> It's my work to get to know what residents do at any time. In which stage they are, what kind of needs they may have, how they organize those needs and to make the right connections. (Female, Community Manager, Coworking Space 2).

In coworking spaces where users were working on a shared project, the role of the community manager was to construct the conditions for positive outcomes. Occasional check-ups and visits to coworkers' offices often accompanied by cookies and little treats were part of their routine. Their aim was to empower start-uppers and early career freelancers to achieve their goals, 'realize their full potential' (Male, Community Manager, Coworking Space 3), to achieve. This type of relational practice lies under what Fletcher (1998) refers to as 'achieving' emphasizing on the 'emotional overlay of situations' (Fletcher, 1998, p. 172).

Likewise, friendly chats and informal introductions are understood as expressions of natural inclinations rather than job tasks or part of formal responsibilities. Those tactics include the adoption of a shouldering behaviour (Fletcher, 1998) which was experienced as part of their way of doing things with an attitude, as Fletcher (1998, p.170) argues, of 'If I don't do it, nobody will'. Indicative is the case of a community manager who narrated to us that during the times he is absent from the coworking space nobody is really talking to each other:

> [I need] to make sure that everybody here talks to each other – this is something I personally and truly want to happen here. (Male, Community Manager, Coworking Space 1)

> I start by introducing myself and asking new members to share with me what is it that they do. It begins informally, it's like we're friends. (Female, Community Manager, Coworking Space 4)

The relational practices described above are characterized by what Sara Ruddick (1989, p.65 in Fletcher, 1998) calls 'preservative love'. This type of love underlines the maternal thinking behind every action which is directed by selfless, and even sometimes, sacrificial giving. As we observed, community managers were prioritizing the maintenance of the coworking community and its wellbeing over their personal lives. Other examples included a community manager who offered to go for after work drinks with someone from the coworking space he thought had troubles in concentrating and structuring his workday. This type of preserving as described by Fletcher (1998) had to do with rescuing or calling attention to problems. In fact, community managers acted as the overly sensitive and compassionate empathizers who incarnated qualities historically attributed to women. Whether it meant to initiate informal chit chats, after work drinks, thank you notes, and birthday surprises to show empathy and care, these activities implied a belief that were part of the innate character of the individual and, thus, initiated effortlessly and naturally.

Due to the lack of overt, tangible control by superiors, their occupation is experienced as relatively autonomous and creative. In this conception, individuals use their imagination, skills, and critical thinking to execute their tasks:

> [...] so, it was like 'hey, you are responsible for this' and then you figure out how you do it [...]. Then [it] might take you a lot, a little? Every time, it varies. This organization [Coworking Space 4] is not micromanaging people. They give you the space to do that your own way. (Male, Host, Coworking Space 3)

The relative autonomy which community managers enjoy renders a series of relational practices, which are crucial for the maintenance of the community, invisible:

> We're playing bad cop, good cop with [name of the community manager]. This is how we manage the unrealistic expectations and demands coworkers have from time to time. (Male, Host, Coworking Space 4)

The above quote gives an example of the various forms of emotional labour (Hesmondhalgh & Baker, 2008; Hochschild, 1979; Veldstra, 2020) community managers have to undertake, while the following depicts the unspeakable aspects of their work:

> It has to do with the more personal aspects of the job […] For example, the other day I had at the office two coworkers, visibly crying, and they came to me for help. They told me that those two coworkers had a serious argument with the other two coworkers and they were seeking help on how to manage this. And I caught myself doing the mediator or even the psychotherapist, the couples' therapist. (Male, Community Manager, Coworking Space 1).

While our interviewees presented a demanding working reality loaded with conflicts that require emotion management, they also emphasized the aspects of their job they personally experienced as positive. In our informal discussions, they repeatedly mentioned the positive feedback they get from coworkers regarding their 'right personality'. Within this context, the self is carefully displayed and performed, and emotion is managed so as not to make coworkers feel overwhelmed. When a coworker was asked about the community manager of the Coworking Space she was based in, she stressed that '[he] is exactly how I like him to be. Not overly friendly, not extremely cold'. Likewise, a community manager argued that success in her role lies at her ability to create her very own conditions for invisibility: '*You know that you've succeeded in your role when your work renders you invisible*'. (Female, Community Manager, Coworking Space 4).

Emerging Forms of Organisation of Work, Long-standing Patterns of (In)visibility

Community managers are an emerging occupational role in coworking spaces. Their work consists of project management of the organisation's internal and external projects, a series of digital marketing tasks to maintain the organisation's networking with external actors and sustain its reputational value, and central components of emotional work, which involve carrying out a series of relational practices within the space and building and sustaining the seemingly serendipitous community. The job incorporates characteristics from the hospitality industry, including multi-tasking, client service, and people management skills. While occupying an explicitly people-oriented position in coworking organisations, their working realities remain largely invisible. This invisibility is reinforced by both the structural informality of coworking spaces and the historical devaluation of relational skills as mere women's capacities.

Our findings suggest that in the coworking spaces under investigation, community managers were informally recruited, under generic job descriptions, if any at all, to mainly execute impression, emotion, and facility management tasks. This informality perpetuates a condition of invisibility for relational practices that are crucial part for the maintenance of the community but experienced as non-work by users, founders, and the people that carry out the work themselves. Thereby, emotion management is more likely to be experienced as part of the self, as part of the manager's 'natural' capacities and personal character, rather than part of the job. The lack of immediate, tangible, control, and direct supervision – long-standing characteristics of low status service work – creates an illusion of autonomy and independence which contributes to the reproduction of invisibility of the majority of relational practices carried out, which are, however, crucial in maintaining the community that has come to be seen as an essential part of coworking.

The self must be carefully displayed to demonstrate what historically have been understood as feminine traits. The caring and warm aspects of their job are difficult to conceptualise by our interviewees as employment skills. Instead, they see their interpersonal and communication skills as part of their natural talent and, as such, experience a sense of self-validation and merit which makes up for the precarious working conditions. However, intimate interactions with coworkers have an exploitative potential, as these 'personal' relations may extend the community managers' working schedules, their job responsibilities, or even their personal boundaries. Staged interactions make the difference in terms of turning a workplace from an indifferent, and transitional office space to a stimulating environment which could hold promise for personal and collective growth. The emerging organisational arrangements of coworking bring to the fore the forms of emotional and relational labour that feminist research has historically fought to make visible and recognised as 'real' work. Despite the fact that within coworking these types of labour take place in the public sphere, the combination of organisational informality of the spaces and feminised connotations inherent in these types of labour means these aspects of the job remain gendered and undervalued as low-paid service work. The hiring practices and short-term contractual character of these jobs mean that interviewees tended to be young, flexible, and without caring responsibilities. When women occupy such positions, they tend to be expected to act as women and thus undertake home care responsibilities. Current recruitment and working practices tend to reinforce a gendered division of labour where coworkers are treated as clients, or even consumers, who conduct 'real work' and community managers are expected to create a collaborative atmosphere through a series of interpersonal gestures that come spontaneously and are part of their 'real self'.

In the corporate world of work, community management tasks would be more likely to be considered as responsibilities of the HR Department. In the flexible organizational set up of coworking, these tasks are partially systematized. Nevertheless, the provision of community management training marks a gradual turn towards the formalisation – if not professionalisation – of the field. Community management roles seem crucial to the maintenance of what coworking literature has repeatedly glorified as its differentiating point. When no community manager is employed, the communitarian aspects of the phenomenon – well documented in the coworking literature – are more likely fluid, informal, if not, entirely absent. The emerging institutional arrangements of the coworking phenomenon shed light to a wide range of relational practices that historically belong to the private, feminine, sphere.

References

Acker, J. (1990). Hierarchies, jobs, bodies: A theory of gendered organizations. *Gender and Society, 4*(2), 139–158. https://doi.org/10.1177/089124390004002002

Akerman, C. (Director). (1975). *Jeanne Dielman, 23 Quai du Commerce, 1080 Bruxelles.*

Avdikos, V., & Iliopoulou, E. (2019). Community-led coworking spaces: From colocation to collaboration and collectivization. In R. Gill, T. Virani, & A. C. Pratt (Eds.), *Creative hubs in question: Place, space and work in the creative economy*. Palgrave Macmillan.

Avdikos, V., & Kalogeresis, A. (2016). Socio-economic profile and working conditions of freelancers in co-working spaces and work collectives: Evidence from the design sector in Greece. *Area, 49*(1), 1–8. https://doi.org/10.1111/area.12279

Blagoev, B., Costas, J., & Karreman, D. (2019). 'We are all herd animals': Community and organizationality in coworking spaces. *Organization, 26*(6), 894–916. https://doi.org/10.1177/1350508418821

Bosma, J. R. (2022). Platformed professionalization: Labor, assets, and earning a livelihood through Airbnb. *Environment and Planning A: Economy and Space, 54*(4), 595–610. https://doi.org/10.1177/0308518X211063492

Conor, B., Gill, R., & Taylor, S. (2015). Gender and creative labour. The Sociological Review, 1–22.

Cox, N., & Federici, S. (1975). *Counter-planning from the kitchen: Wages for housework, a perspective on capital and the left*. New York Wages for Housework Committee and Falling Wall Press. Retrieved March 7, 2022, from https://monoskop.org/images/0/03/Cox_Nicole_Federici_Silvia_Counter-Planning_from_the_Kitchen_Wages_for_Housework_A_Perspective_on_Capital_and_the_Left.pdf

Dalla Costa, M., & James, S. (1972). *Women and the subversion of the community*. Retrieved March 7, 2022, from https://files.libcom.org/files/Dalla%20Costa%20and%20James%20-%20Women%20and%20the%20Subversion%20of%20the%20Community.pdf

Daniels, A. K. (1987). Invisible work. *Social Problems, 34*(5), 403–415. https://doi.org/10.2307/800538

de Peuter, G., Cohen, S. N., & Saraco, F. (2017). The ambivalence of coworking: On the politics of an emerging work practice. *European Journal of Cultural Studies, 20*(6), 687–706. https://doi.org/10.1177/136754941773299

Duffy, M. (2007). Doing the dirty work: Gender, race, and reproductive labor in historical perspective. *Gender and Society, 21*(3), 313–336. https://doi.org/10.1177/0891243207300764

Dunn, M. (2020). Making gigs work: Digital platforms, job quality and worker motivations. *New Technology, Work and Employment, 35*(2), 232–249. https://doi.org/10.1111/ntwe.12167

Federici, S. (2006). *Precarious labor: A feminist viewpoint*. Retrieved March 7, 2022, from https://inthemiddleofthewhirlwind.wordpress.com/precarious-labor-a-feminist-viewpoint/

Ferrant, G., Pesando, L. M., & Nowaska, K. (2014). *Unpaid care work: The missing link in the analysis of gender gaps in labour outcomes*. Retrieved March 20, 2022, from https://www.oecd.org/dev/development-gender/Unpaid:care_work.pdf

Fletcher, J. K. (1998). Relational practice: A feminist reconstruction of work. *Journal of Management Inquiry, 7*(2), 163–186. https://doi.org/10.1177/105649269872012

Gill, R. (2007). Technobohemians or the new Cybertariat? New media work in Amsterdam a decade after the Web [Report prepared by Rosalind Gill for the Institute of Network Cultures]. Institute of Network Cultures.

Gregg, M., & Lodato, T. (2018). Managing community: Coworking, hospitality and the future of work. In B. Röttger-Rössler & J. Slaby (Eds.), *Affect in relation. Families, places, technologies* (pp. 175–196). Routledge.
Hesmondhalgh, D., & Baker, S. (2008). Creative work and emotional labour in the television industry. *Theory, Culture and Society, 25*(97). https://doi.org/10.1177/0263276408097798
Hochschild, A. R. (1979). Emotion work, feeling rules, and social structure. *American Journal of Sociology, 85*(3), 551–575. https://www.jstor.org/stable/2778583
Huws, U., Spencer, N. H., Syrdal, D. S., & Holts, K. (2018). *Work in the European gig economy* (FEPS – Foundation for European progressive studies). Europe and University of Hertfordshire.
McDowell, L. (2009). *Working bodies: Interactive service employment and workplace identities.* Wiley-Blackwell.
Merkel, J. (2015). Coworking in the city. *Ephemera, 15*(1), 121–139.
Merkel, J. (2019a). Curating strangers. In R. Gill, A. C. Pratt, & T. Virani (Eds.), *Creative hubs in question* (1st ed., pp. 51–68). Palgrave Macmillan.
Merkel, J. (2019b). 'Freelance isn't free.' Co-working as a critical urban practice to cope with informality in creative labour markets. *Urban Studies, 56*(3), 526–547. https://doi.org/10.1177/0042098018782374
Moriset, B. (2014). *Building new places of the creative economy. The rise of coworking spaces.* In 2nd geography of innovation, international conference 2014. Utrecht University, Utrecht (The Netherlands). Retrieved March 20, 2022, from https://halshs.archives-ouvertes.fr/halshs-00914075/document
Mould, O., Vorley, T., & Liu, K. (2014). Invisible creativity? Highlighting the hidden impact of freelancing in London's creative industries. *European Planning Studies, 22*(12), 2436–2455. https://doi.org/10.1080/09654313.2013.790587
Pacchi, C., & Mariotti, I. (2021). Shared spaces or shelters for precarious workers? Coworking spaces in Italy. *Professions & Professionalism, 11*(1). https://doi.org/10.7577/pp.3911
Papageorgiou, A., Michailidou, Martha, Avdikos, V., & Merkel, J. (2022). *Policy brief 2: The impact of coworking spaces on coworkers' well-being and skills' development.* Panteion University. Retrieved February 23, 2023, from https://www.researchgate.net/publication/366086637_Policy_Brief_2_The_impact_of_coworking_spaces_on_coworkers'_well-being_and_skills'_development
Prassl, J. (2018). *Humans as a service: The promise and perils of work in the gig economy.* University of Oxford.
Reuschke, D., Clifton, N., & Fisher, M. (2021). Coworking in homes – Mitigating the tensions of the freelance economy. *Geoforum, 119*, 122–132. https://doi.org/10.1016/j.geoforum.2021.01.005
Ruddick, S. (1989). Maternal thinking: Towards a politics of peace (Vol. 17, Issue 2, pp. 85–104). The Women's Press.
Veldstra, C. (2020). Bad feeling at work: Emotional labour, precarity, and the affective economy. *Cultural Studies, 34*(1), 1–24. https://doi.org/10.1080/09502386.2018.1555269
Virani, T., & Gill, R. (2019). Hip hub? Class, race and gender in creative hubs. In R. Gill, A. C. Pratt, & T. Virani (Eds.), *Creative hubs in question: Place, space and work in the creative economy.* Palgrave Macmillan.
Webster, A. N., & Zhang, Q. (2020). Careers delivered from the kitchen? Immigrant women small-scale entrepreneurs working in the growing Nordic platform economy. *NORA – Nordic Journal of Feminist and Gender Research, 28*(2), 113–125. https://doi.org/10.1080/08038740.2020.1714725
Wittel, A. (2001). Toward a network sociality. *Theory, Culture and Society, 18*(6), 51–76. https://doi.org/10.1177/02632760101800600
Worth, N., & Karaagac, A. E. (2021). Accounting for absences and ambiguities in the freelancing labour relation. *Tijdschrift Voor Economische En Sociale Geografie*, 1–13. https://doi.org/10.1111/tesg.12491

Exhausting Coworking: On the Implications of Reproductive Work for Coworkers' Subjectivities

Nikos Gatsinos and Malte Höfner

Introduction

Coworking is regarded as another way of organising work between working in a traditional workplace or working in isolation at home (Gandini, 2015), in a space where freelancers enjoy the sociality of 'working alone together' (Spinuzzi, 2012), and where the opportunity for innovative collaboration between microbusinesses may be cultivated (Capdevila, 2013; Nakano et al., 2020). The proliferation of coworking spaces is evident especially since the global financial and economic crisis of 2007/2008 (Gandini, 2015; Merkel, 2019a), which echoes its characterisation as a 'worker-developed response' (de Peuter et al., 2017, p. 689), or as 'a practice of self-help and self-organisation' (Merkel, 2019b, p. 540) against precariousness associated with creative and cultural labour (Genders, 2022; Gill & Pratt, 2008; McRobbie, 2009). Therefore, one can consider coworking spaces as 'new ways of organising labour [that] enabl[e] mutual support amongst freelancers and self-employed persons' (Merkel, 2015, p. 122).

Collaborative workspaces (hereinafter 'CWS') refer to a variety of shared workplaces (Schmidt & Brinks, 2017), and differ substantially in their organisational forms and functions. Avdikos and Iliopoulou distinguish community-led from entrepreneurial-led coworking spaces, highlighting that community-led coworking spaces (hereinafter 'CLCS') 'are usually not-for-profit ventures that accommodate (in semi-permanent terms) a small number of creatives with a kind of social and even political proximity between them' (2019, p. 112). The distinction between community-led and entrepreneurial-led spaces is not to be considered as absolute and/or rigid, as CWS constantly oscillate between the commons and the market (Avdikos & Pettas, 2021). Nevertheless, the typology of CLCS provided by Avdikos

N. Gatsinos (✉) · M. Höfner
University of Graz, Department of Geography and Regional Science, Graz, Austria
e-mail: nikolaos.gatsinos@uni-graz.at; malte.hoefner@uni-graz.at

J. Merkel et al. (eds.), *Coworking Spaces*, https://doi.org/10.1007/978-3-031-42268-3_9

and Iliopoulou (2019) is fruitful for our work, as we investigated a grassroots CWS (Gandini & Cossu, 2021) in the medium-sized city of Graz, Austria, which can be characterised as CLCS due to the not-for-profit character of the space, the ideological commitment of its members, as well as the collegial atmosphere that prevails within the space (Avdikos & Iliopoulou, 2019, p. 125).

What motivated our work were critical studies pointing to the various ambivalences that accompany coworking (de Peuter et al., 2017; Waters-Lynch & Duff, 2021). With that in mind, we followed professionals who used to cowork in the past, to explore whether they quit working in their CWS (at least partly) driven from an ambivalence they felt towards coworking. Conducting follow-up interviews, we saw that their ambivalence largely stemmed from the disproportionate reproductive work they had to provide for the community and for the CLCS, which led them to being exhausted. While we acknowledge the compelling argument that coworking spaces may function as reproducers of a neoliberal subjectivity (Bandinelli, 2020) in a context of prevalent 'collaborative individualism' (Bandinelli & Gandini, 2019, p. 103), in our work we came across how 'social relationships formed in coworking spaces influence coworkers' in a way that 'subject positions might change through co-working' (Merkel, 2019b, p. 540). In this vein, we looked at 'ordinary practices' (Holmes, 2018) such as sharing, caring, and curating work that take place within community-led coworking, to see whether opportunities for 'facilitat[ing] mutual awareness and support' (Merkel, 2019b, p. 531) are cultivated. In this chapter, we first reflect on the ambivalences of coworking as they are portrayed in literature. Then, we focus on the ambivalence driven by exhaustion, and explore the diverse forms of work necessary to maintain a CLCS. To highlight how reproduction is sustained through work, we apply a diverse economies' framework (Gibson-Graham, 2008). In the final section, we provide food for thought on how exhaustion caused by reproductive work (Resch et al., 2021) may serve as a useful lens, when read against the backdrop of an alleged neoliberal subjectivity that is imputed to many CWS.

Coworking in Ambivalence

The work of de Peuter et al. (2017) provides a strong account of the contradictions emerging from coworking practice. Firstly, they contend that, although coworking 'emerged from below', it is now 'commodified, increasingly corporatised and ultimately facilitates the flexibilisation of labour' (ibid., p. 689). Moreover, they suggest that coworking entails the 'tensions between coworking's counter-corporate identity and its recapitulation of neoliberal norms' (ibid., p. 689). In similar fashion, Bandinelli (2020) argues that the ambivalence of coworking lies in the contradiction between the 'instrumental logic of the entrepreneur of the self' and an 'attempt to produce the conditions for solidarity' (ibid., p. 7). In contrast, Waters-Lynch and Duff (2021) argue that ambivalence about coworking and the irregularities of freelance work are two distinct phenomena. More specifically, it is '[c]oworking itself, rather than [...] the broader conditions of precarity' that is causally related to the ambivalence (ibid., p. 385). Said alternatively, Waters-Lynch and Duff (2021) do not regard the ambivalence as 'intrinsic to the culture of coworking' (de Peuter et al., 2017, p. 689), stemming from the insecurities of freelance work; rather, they perceive it as based on the conflicts arising between the user-generated affective atmospheres and their private capture by firms running the CWS.

Even though we find the latter compelling and befitting the picture of entrepreneurial-led coworking spaces, we argue it is critical to take a step back and see how it applies to CLCS. Waters-Lynch and Duff (2021) and Bernhardt (2021) argue that the affective atmospheres in CWS are (re)produced by their participants. The 'senses of community' (Garrett et al., 2017) and of mutual support are part of these atmospheres and constitute a collectively (re)produced good. However, instead of this good being consumed by the members, it is extracted as added value from the company and then commodified as a service. Contrarily, in CLCS there is rarely one single enterprise that operates the space and claims the affective atmospheres as a commodified product.[1] These spaces are mostly run by its members; therefore, the 'company' is constituted by the members themselves. Insofar as a coworker joins a CLCS, they need to partake in the work needed for sustaining the space, taking part in the (re)production and consumption of its affective atmospheres. Simply put, in a CLCS there is no distinction between the producers and the consumers of an affective atmosphere. It is exactly at this point where Waters-Lynch and Duff's (2021) argument becomes more nuanced and useful for our research: drawing on the commons' theory, they write about the need to take part in the (re)production and consumption of a common, pooled resource, and highlight the need to explicitly address the principle of common organisation around these processes. Alternatively stated, the commons need to be governed as such (Ostrom, 1990). Understanding the importance of governing the affective atmospheres and the organisation of reproductive work is what we think is missing from the CLCS we observed.

The Ambivalence Stemming from Exhaustion

Our research sample consists of professionals who used to practise community-led coworking in the past, but have since left community-driven working arrangements. They were originally interviewed by Malte Höfner in the years 2015/2016 (at a time when they used a CLCS). Follow-up interviews were conducted by both authors in late 2021 and the beginning of 2023 (when they had stopped coworking in a community). The aim of these interviews was to facilitate participants' reflections on their past coworking experiences and to portray how community-led coworking related to their professional and personal life trajectories. Our cases refer to a particular

[1] For a critical view on the matter, see Arvidsson (2018), who contends that practising community in coworking may be part of branding the imaginary of community ('social branding').

workspace that we have given the name 'Creation Mill', which is known for its openness to many forms of collaboration, supporting interdisciplinary exchange, initiatives, and diversity. This means that it is not exclusively devoted to income-generating forms of labour. People use (or have used) this space for self-development and experimental side-projects, hobbies, and businesses. Hence, in most cases, the profiles of the coworkers are multi-faceted. That means, in addition to their activity in Creation Mill, some also carry out bread-and-butter jobs elsewhere to make a living but follow their side-projects in their 'free time' at Creation Mill. However, paid work that is carried out at Creation Mill is the singular source of income for others. During his active phase of coworking, Moritz (in parallel to a short-term contract as a university lecturer) even referred to the space as being something more than coworking; namely, being a 'space of possibilities' (2015, 06:34). Interestingly, the coworkers' positive feelings towards community-led coworking explicitly mentioned in the past (2015) still echoed when they were asked again during the follow-up interviews (2021/2023):

> I was in Creative Mill with seven to nine others in one room. It was a space – not only for creatives. For all people who needed a space and not to work from their home, to do work or to fill it with things they wanted to do. I liked it a lot, this spirit. And it's cool that it exists. That it still exists. (Sophie, 2023, 03:25)

> [...], it was a huge experiment and just so much fun. Of course, it was about somehow (laughs) to have the rent and maybe somehow money to eat, but the most important thing was [...] the realisation, to try things out, to do interesting things, [...]. (Caro, 2021, 49:22)

Sophie and Caro both describe their former time at Creation Mill as something special and valuable; they still seem emotionally attached to the space even though they are no longer there and appreciate that such a place still exists. The encounters people had at Creation Mill become infused with meaning through (authentic) sharing of common values, best described as 'being-in-common' (Smith, 2020, p. 593) in search for new post-capitalist work practices. A sense of belongingness was crafted through being 'together' (see Moritz) 'viscerally and affectively empowering alternative[s]' (ibid., p. 596). Given the enjoyment they found in their coworking community, we asked: what was the reason they decided to quit? Did they face an ambivalence against coworking, pushing them out at a certain point?

Throughout the interviews we identified two intertwined reasons why our coworkers decided to distance themselves from the CLCS: the first was the enormous amount of reproductive work the coworkers had to carry out, crucial for sustaining the affective atmospheres as the core of the space. The second reason relates to their need to grow as freelancers as well as the time demanded by their families (Caro, Moritz, Sophie), which brought even more stress to their everyday workload.

Community-led Coworking Needs Work

Producing and sustaining affective spatial 'atmospheres of communitarisation' (Bernhardt, 2021, p. 231 f., own translation) or an 'ambience of collaboration' (Orel & Almeida, 2019, p. 274) within a CWS, is the cornerstone of coworking.

The curatorial work necessary for the reproduction of this working environment is usually carried out by the host: '[the host has a] crucial role in facilitating coworking as a collaborative approach and translating coworking values into the space' (Merkel, 2015, p. 125). In other words, interaction and collaboration between coworkers are not to be taken for granted (Spinuzzi, 2012), as their facilitation needs to be implemented by a gatekeeper, usually either the proprietor of the space or a paid community manager (Merkel, 2015). In a CLCS however, there is no owner, as CLCS are usually brought together by the members, who share the costs of the space as well as the work required for its reproduction. In the CLCS we observed, the rent and other costs related to material infrastructure were indeed equally shared among the members. What remains is the work necessary for the reproduction of the affective atmospheres and how this work is distributed among the members.

Curating the CLCS as places of creativity (Brown, 2017) and spaces of possibility is an essential component of the work necessary to reproduce the space, including its embeddedness into the local cityscape.[2] In a CLCS, a host's job is distributed to more than one member:

> For example, when the crime novelist [a member] is finished [with his book], then the novel is exhibited [in the display]. Or when the flood was with the high tide of the [river] Mur, there [display] was this driftwood [they put it there to show]. Or [the] old bread [used for an art project] – when the sun shone in. [...] the 'city simply pulls itself in' there and becomes more concentrated [...]. In other places, it has more of an office character; we share an office. (Moritz, 2015, 10:08)

Sophie, who was the host at that time, showed her acknowledgement to others' curatorial work (appreciating Moritz's playful way of exhibiting the public display). On the one hand, Creation Mill is perceived as a space of possibilities, as Moritz, Sophie, and Caro initially described it above. On the other hand, this space of possibilities exists as such only if it is taken care of, as expressed by Sophie. This requires not only time, but also energy, and a certain level of commitment from the individual. Though the reproductive work is indeed distributed to more than one member, one can observe instances of uneven distribution of such work. Sophie recognised there are some members who are more 'silent' and 'just' want to use their desk, seeming to have less interest in the social reproduction of the space. This shows imbalances within the areas of responsibility:

> Yes, there are some who care about the kitchen. There are some for whom the displays are important, more important. There are some who just care about their own desk [...] and they don't need anything else and don't want to put any energy into writing something onto the window, for example. (Sophie, 2015, 32)

Not surprisingly, practices of nurturing or care within coworking spaces are related to 'mothering' and 'mutualism' (Belk, 2010; see also Merkel, 2015, p. 128),

---

[2] These tasks may sound trivial, but we know from private conversations with the current host that these things make pedestrians curious; they stop and wonder what is happening there and what the [CLCS] actually is about.

since they are not necessarily connected with demands for reciprocation. Something Sophie (2023, 27:46) today strongly identifies with when reflecting on her time as the host of Creation Mill. Hence, the term 'host' becomes controversial within a CLCS. According to the principles of the CLCS we observed, the work needed to maintain the space should be equally distributed among the members; but as suggested above, a host may carry more responsibilities, which in turn intensifies struggles when coupled with an increased workload (see Section "'She actually did a lot. and then, it was too much [...]': The limits of (exhausting) reproductive work in CLCS").

Applying a diverse economies' framework (Gibson-Graham, 2008), the above mentioned practices of reproducing the community space can be seen as forms of work (or 'labour', in the diverse economies' vocabulary) that may be infused with reciprocity, exchange, and mutuality. In a series of critical research in human geography, Gibson-Graham (2006, 2008, 2014) have provided us with a different mapping of economy that expands the (often narrow) Marxist analytical lenses of the economy. Adopting a thick description approach, they point to the variety of diverse economies performed simultaneously within, apart from, and as part of capitalist economies. However, instead of solely focusing on 'capitalocentric' logics and monetary transactions, Gibson-Graham research the heterogeneity of diverse economies. They contend that the economy consists of capitalist, alternative capitalist and non-capitalist enterprises, labour, property, transactions, and finance (2014, p. 150). Morrow and Dombroski (2015) draw on the same theoretical framework and focus specifically on diverse forms of labour performed in communities. In their contribution, they distinguish between three forms of labour activity, namely provisioning, income generating, and caring (ibid., p. 85, see also Holmes, 2018). They sub-categorise these forms of labour activity regarding how they are compensated (or not) (waged, alternatively paid, unpaid).

In our case study, we see that work necessary to reproduce the CLCS and their affective atmospheres takes the form of either unpaid or alternatively paid labour. In particular, there are coworkers who disproportionately assume caring activities for the space (see Sophie's statement above), which, through a diverse economies' lens, refer to provisioning (e.g., preserving, making, repairing, and DIY labour; see also Holmes, 2018; Smith, 2020) and caring (extra care work performed in workspaces, caring for community spaces[3]) (Morrow & Dombroski, 2015, p. 85). In our sample, there are also cases when this labour assumes a reciprocal character, where the (reproductive) coworker is alternatively paid (e.g., in-kind payments for provisioning, extra care work that is only partially compensated through communal labour carried out by others; see previous statement) (ibid., p. 85). It is a diverse economies' lens that gives 'the ordinary work of everyday life' (Holmes, 2018, p. 145) another level of reading; as *ordinary but not self-evident*.

[3] Something subsumed as a 'post-capitalist practice' by Smith (2020).

'She Actually Did a Lot. And Then, It Was Too Much [...]'[4]: The Limits of (Exhausting) Reproductive Work in CLCS

Our interviewees demonstrate that, initially, coworkers enjoyed the unconstrained experimentation taking place within a tight-knit community; they felt part of a vibrant community involved in the neighbourhood, as well as in the local cultural and arts scene. Similar to Resch et al. (2021), who suggest that there are cases when 'authentic sharing is more important for successful collaboration than just professional content' (ibid., p. 799), our coworkers did not prioritise income-earning over meaningful work (Caro, 2021; Moritz, 2015), and they indeed found meaning in the potential of 'changing the world' while being part of the community (Sophie, 2015). This is best highlighted by the pleasure found within a 'tribal fantasy' of 'belonging' (Resch et al., 2021, p. 799, p. 788). As Sophie vividly describes, in 2015 she so closely identified with the community that she saw her coworkers as 'world changers':

> [...] on the one hand, there are certainly many of us in precarious jobs and it all seems to us, but not too tragic, because we all have each other. [...]. If someone gave me a mission to save the world, I know I'd ask three [people] in my coworking space – they can do it [laughs!]. That's already worth a lot. [...]. (Sophie, 2015, 10)

Our cases mirror what is described as the 'painful pleasure' in a 'tribal fantasy' (ibid., p. 799). Resch et al. (2021) suggest that coworkers initially experience the excessive pleasure of identifying with a community centred on a communal (imagined) purpose. This is in turn performed through practices of authentic sharing (i.e., driven by intrinsic motivation and not as a strategy for entrepreneurial success, or for reducing costs), which point to 'seeking purposeful wholeness' in community (ibid., p. 795). At a later stage though, acts of 'sharing the social' in pursuit of belonging may lead to emotional exhaustion and can become 'finite' due to scarce resources like time (Kennedy, 2016, p. 461 f.). In a similar manner, our interviewees' retrospective reflections resonate with what Resch et al. term as the 'pain of excess' (2021, p. 800):

> [...] I was born in a mountain village, and I felt the same sitting in [...] Creation Mill. There is a glass front you know. So, people stepping by, they knew me, I knew them, so they entered. It happened...every working day [...]. I loved this conversation[s] that happened but it was...I felt "okay, now it's time to reduce a bit, to focus, to make [everything] a little bit smaller [...]." [...]. Then I realised, okay, no I have to go now, I have to change, I have to make a change. (Sophie, 2023, 03:25)

Here we see that when the capacity to care for yourself while taking care of others becomes exhausted (see also Höfner & Gatsinos, 2023), members might try to loosen their connections with the community in an attempt to dissolve this tension (Resch et al., 2021). 'The painful edge of this [is] observed in the course of informal

[4] Quote from former Coworker Caro (2021, 29:38) referring to the founder and first 'host' of Creation Mill.

care practices, where people prioriti[se] care for others over self-care. This le[ads] to the exhaustion of a group of highly engaged members' (ibid., p. 800). Caro told us about the founder, who, after slowly stepping away from self-employment and leaving Creation Mill due to professional change, still tried to maintain the connections to the space by fulfilling former host's 'duties':

> [S]he came to clean the coffee machine, did the bills, and made sure that everything fitted and organised the cleaning lady somehow. She actually did a lot. And then it was too much for her, for the fact that she could no longer use [Creation Mill] herself. (Caro, 2021, 29:38)

Despite the strong commitment, this didn't work out given her new responsibilities and other obligations outside the space (e.g., family, new job). Particularly enlightening is the observation that most community members (except for silent members, i.e., coworkers who might either refrain from collectively carried out curatorial practices or partake only if explicitly asked for) realise there is a lot of work to be done, and they are also aware that this often has to do with the character of the person:

> Of course, there are people, such as Sophie, who are here a lot and have a good sense of what's going on [...]. That means that it also depends on the person, that some people feel more responsible because they are here more often or because of their type [character], which is of course often agonising because other people simply don't see the things that you see, and you don't want to offend anyone and that is a very thin line and often also exhausting. There are also conflicts, but of course these are 'adult', we are grown-ups. [...]. (Moritz, 2015, 25:12)

However, the knowledge about these structures in a working community does not seem to have significant impacts on how the (unevenly) distributed tasks of the workspace's care work could be rearranged or transformed. On the contrary, it even revealed a lack of clarity between coworkers on how to carry out common responsibilities in shared spaces of work:

> [...] perhaps one person [referring to the host] had the power to admonish the others if they didn't pay [the rent] [...]. So, most things have been decided collectively. And I think many of those who ['just'] had a desk, perhaps didn't even know who was actually in charge of the lease, who was actually doing the contractual stuff in the first place. [...]. (Caro, 2021, 23:17)

The fact that all important decisions are supposed to be made jointly does not mean that all persons share the same degree of responsibility. We know from both Caro and the current host that legal matters (rent, insurance, liability, bank account, etc.) are still – at least organisationally – the responsibility of the host. And, albeit the 'silent members' 'don't see' (Moritz) many forms of care work and responsibilities, they are still somehow acknowledged since they partake in maintaining the space financially (Caro mentioned this point when struggling for new members, and the current host has confessed the importance of the 'silent members' when membership rates were too low to finance the space). The exhaustive character of taking care is therefore exacerbated by an uneven distribution of reproductive tasks necessary for the space, which gives rise to a dichotomy between the 'carers' and the 'silent members'.

Discussing Exhaustion in CLCS: Implications For Neoliberal Subjectivities or a 'Coming Community'?

In our research we saw that at some point coworkers were torn between (a) a need to grow as professionals and/or for more disposable time on the one hand, and (b) demands for non-remunerated, time- and energy-consuming reproductive work in the CLCS on the other. Following critical studies about the pervasion of neoliberalism in CWS, Merkel (2019b) suggests that, although important, '[t]his approach leaves little theoretical scope for workers' agency and conceptualises workers' subject position as only constituted through their work' (ibid., p. 540). Following our ex-coworkers' reflective discussions about their time practising work in a community, we faced this precise question: Does the coworkers' exit from the CLCS underpin the narrative of a so-called neoliberal subjectivity, of a self-entrepreneur who withdraws to isolation trying to accommodate their interests in an individualistic manner? Or may community-led coworking be also seen as a collective practice of working exceeding the reproduction of neoliberal subjects (Sandoval, 2016)?

Our work produces interesting insights when placed in conversation with Johns and Hall's (2020) scholarship on sharing in a FabLab in Manchester. They similarly point to the 'obstacles to sharing' in a collaborative workspace (ibid., p. 1306 f.). In their case, they witnessed individualist logics at play in performing collabora tion within the CWS, where the main driver for 'sharing' was the pursuit of profit. The ideals of the FabLab they researched referred to makers exiting the space after their professional consolidation. In that sense, it was not genuine altruism that produced the sharing practices, but commercialisation, which was '[...] seen as the beginning of the end of the makers' relationship with each other and the FabLab [...]' (Johns & Hall, 2020, p.1308). Here, it seems that the makers used the space strategically; once they reached a point to take off professionally, they decreased the degree of sharing. In contrast, in our case, we observed that people had to decrease their involvement in sharing with the community not because they *already* 'grew' as professionals, and thus *did not need* to share anymore, but *in order* to grow, as well as *in order* to afford the time needed for new responsibilities of their growing families. In other words, our cases do not illustrate the successful exit of an established professional out of the CLCS as an objective; on the contrary, people were rather torn emotionally about the decision to drop out. We saw people who *wanted* to stay but felt they *could not.*

It is instances like this which we believe should not go lost under a narrative of a neoliberal subjectivity instrumentalising sharing practices for the sake of their own development (e.g., Bandinelli, 2020; Bandinelli & Gandini, 2019; de Peuter et al., 2017). The fact that coworkers felt emotionally challenged before exiting the space, and that they still appreciate its existence and community values, indicate they genuinely wanted to practise sharing but were challenged by increasingly demanding professional and personal life arrangements. Moreover, contradictory forces might simultaneously influence coworkers' agency; on the one hand there is their genuine willingness to share in the context of a gift culture and/or reciprocity (Belk, 2010), and a push towards reducing the sharing activities because of restricted resources, on the other. Authentic sharing practices based on reciprocity may coexist with the intensities and structural boundaries a capitalocentric economy imposes on freelancers: 'values, such as cooperation, altruism, generosity, mutuality, solidarity [...] pervade the everyday circuits of value in parallel with anxiety and fear of antagonism' (Gritzas & Kavoulakos, 2016, p. 930). Therefore, we think that research on coworking should be attentive to practises that 'hold the possibility for altering, undermining, and undoing [capitalist and exploitative] relations', that is to say, 'for making new subjects' (Mitchell et al., 2003, p. 433). Given the community's (emotional) ascription to a common imaginary and its orientation to the future, we would say we saw a 'coming community', to use Arvidsson's (2018, p. 295) interpretation of Agamben (1993).

Conclusion

In our chapter we focused on the reasons why coworkers of a CLCS decided to leave their community. The aim was to juxtapose these reasons with the common scholarly narrative about the prevalence of neoliberal subjectivity in CWS. Our interviewees presented a picture that rather deviates from this narrative. What led them to the exit was not their will to fulfil their self-interest as entrepreneurs (Bandinelli, 2020). Rather, coping with the tremendous amount of reproductive work they had to carry out for the CLCS gradually became a burden. The community demanded time and energy from the coworkers, resources that were also asked for from their personal and professional arrangements. The fact that coworkers abandoned their former CLCS due to exhaustion, we suggest, offers interesting insights regarding the coworkers' subjectivity and how it is nurtured within CLCS.

Following Herod, we acknowledge that workers have the incentive to 'shape the economic landscape in ways that facilitate this self-reproduction' (2001, p. 33); that is observed when coworkers talk of the tensions arising within a CLCS in an emotionally charged way. They want to share and shape the spaces of (co-)production into spaces of their reproduction in ways they feel comfortable with (Richardson, 2016). The fact that coworkers *feel* for the CLCS is proof that they perceive them as more-than-labouring spaces, as spaces where they perform much more than just 'working alone together' (Spinuzzi, 2012); they see themselves as social entities 'co-constructing community' (Garrett et al., 2017) and 'curating the[ir] third place' (Brown, 2017). As Arvidsson (2018) suggests, we did not encounter a community in the traditional sense (i.e., connected by preceding social bonds) but rather a community of practice that is bound together by a collective ethos and a future-oriented imaginary. That, in turn, is why we see potential in community-led coworking. For this potential to be materialised, there needs to be a structure of 'equitable governance' over the reproductive work that takes place. What is needed in our perspective, is a change in the discourse around what work is within CLCS and who performs it, how it is distributed, and what is or is not shared among the members, so that its diversity is explicitly addressed and equally met by as many members as possible. As Waters-Lynch and Duff (2021) address the necessity to collectively organise the governance of an affective commons within a coworking space, members/coworkers might need to explicitly address diverse forms of work as prerequisites for the(ir) community-led spaces' social reproduction. This, in turn, might reveal the amount of exploitation of (unevenly assumed) reproductive work within the community, insofar as one could learn to practise and experiment with different forms of work within the community in the sense of a real unconstrained experimentation (Vidaillet & Bousalham, 2020).

References

Agamben, G. (1993). *The coming community*. University of Minnesota Press.

Arvidsson, A. (2018). Value and virtue in the sharing economy. *The Sociological Review, 66*(2), 289–301. https://doi.org/10.1177/0038026118758531

Avdikos, V., & Iliopoulou, E. (2019). Community-led coworking spaces: From co-location to collaboration and collectivization. In R. Gill, A. C. Pratt, & T. E. Virani (Eds.), *Creative hubs in question. Dynamics of virtual work* (pp. 111–129). Palgrave Macmillan. https://doi.org/10.1007/978-3-030-10653-9_6

Avdikos, V., & Pettas, D. (2021). The new topologies of collaborative workspace assemblages between the market and the commons. *Geoforum, 121*, 44–52. https://doi.org/10.1016/j.geoforum.2021.02.012

Bandinelli, C. (2020). The production of subjectivity in neoliberal culture industries: The case of coworking spaces. *International Journal of Cultural Studies, 23*(1), 3–19. https://doi.org/10.1177/1367877919878449

Bandinelli, C., & Gandini, A. (2019). Hubs vs Networks in the creative economy: towards a 'collaborative individualism'. In R. Gill, A. C. Pratt, & T. E. Virani (Eds.), *Creative hubs in question. Dynamics of virtual work* (pp. 89–110). Palgrave Macmillan. https://doi.org/10.1007/978-3-030-10653-9_5

Belk, R. (2010). Sharing. *Journal of Consumer Research, 36*(5), 715–734. https://doi.org/10.1086/612649

Bernhardt, A. (2021). *Coworking-Atmosphären: Zum Zusammenspiel von kuratierten Räumen und der Sicht der Coworkenden als raumhandelnde Subjekte*. Springer Fachmedien. https://doi.org/10.1007/978-3-658-35888-4

Brown, J. (2017). Curating the "Third Place"? Coworking and the mediation of creativity. *Geoforum, 82*, 112–126. https://doi.org/10.1016/j.geoforum.2017.04.006

Capdevila, I. (2013, December 9). *Knowledge dynamics in localised communities: Coworking spaces as microclusters*. SSRN. Retrieved March 1, 2023, from http://ssrn.com/abstract=2414121

de Peuter, G., Cohen, N. S., & Saraco, F. (2017). The ambivalence of coworking: On the politics of an emerging work practice. *European Journal of Cultural Studies, 20*(6), 687–706. https://doi.org/10.1177/1367549417732997

Gandini, A. (2015). The rise of coworking spaces: A literature review*. In P. Bialski, H. Derwanz, B. Otto, & H. Vollmer (Eds.), *'Saving' the city: Collective low-budget organising and urban practice* (Ephemera, 15(1), pp. 193–205). Mayflybooks. https://ephemerajournal.org/contribution/rise-coworking-spaces-literature-review

Gandini, A., & Cossu, A. (2021). The third wave of coworking: 'Neo-corporate' model versus 'resilient' practice. *European Journal of Cultural Studies, 24*(2), 430–447. https://doi.org/10.1177/1367549419886060

Garrett, L. E., Spreitzer, G. M., & Bacevice, P. A. (2017). Co-constructing a sense of community at work: The emergence of community in coworking spaces. *Organization Studies, 38*(6), 821–842. https://doi.org/10.1177/0170840616685354

Genders, A. (2022). Precarious work and creative placemaking: Freelance labour in Bristol. *Cultural Trends, 31*(5), 433–447. https://doi.org/10.1080/09548963.2021.2009735

Gibson-Graham, J. K. (2006). *The end of capitalism (as we knew it): A feminist critique of political economy; with a new introduction*. University of Minnesota Press.

Gibson-Graham, J. K. (2008). Diverse economies: Performative practices for 'other worlds'. *Progress in Human Geography, 32*(5), 613–632. https://doi.org/10.1177/0309132508090821

Gibson-Graham, J. K. (2014). Rethinking the economy with thick description and weak theory. *Current Anthropology, 55*(9), 147–153. https://doi.org/10.1086/676646

Gill, R., & Pratt, A. (2008). In the social factory? Immaterial labour, precariousness and cultural work. *Theory, Culture & Society, 25*(7–8), 1–30. https://doi.org/10.1177/0263276408097794

Gritzas, G., & Kavoulakos, K. I. (2016). Diverse economies and alternative spaces: An overview of approaches and practices. *European Urban and Regional Studies, 23*(4), 917–934. https://doi.org/10.1177/0969776415573778

Herod, A. (2001). *Labor geographies: Workers and the landscapes of capitalism*. Guilford Press.

Höfner, M., & Gatsinos, N. (2023). Unfinished sympathy: On the limitations of sharing as a work practice in community-led coworking. Cogent Social Sciences, 9(2), 2245236. https://doi.org/10.1080/23311886.2023.2245236

Holmes, H. (2018). New spaces, ordinary practices: Circulating and sharing within diverse economies of provisioning. *Geoforum, 88*, 138–147. https://doi.org/10.1016/j.geoforum.2017.11.022

Johns, J., & Hall, S. M. (2020). 'I have so little time […] I got shit I need to do': Critical perspectives on making and sharing in Manchester's FabLab. *Environment and Planning A: Economy and Space, 52*(7), 1292–1312. https://doi.org/10.1177/0308518X19897918

Kennedy, J. (2016). Conceptual boundaries of sharing. *Information, Communication & Society, 19*(4), 461–474. https://doi.org/10.1080/1369118X.2015.1046894

McRobbie, A. (2009). Reflections on precarious work in the cultural sector. In B. Lange, A. Kalandides, B. Stöber, & I. Wellmann (Eds.), *Governance der Kreativwirtschaft. Diagnosen und Handlungsoptionen* (pp. 123–138). transcript Verlag. https://doi.org/10.1515/9783839409961-007

Merkel, J. (2015). Coworking in the city. In P. Bialski, H. Derwanz, B. Otto & H. Vollmer (Eds.), *'Saving' the city: Collective low-budget organising and urban practice* (Ephemera, 15(1), pp. 121–139). Mayflybooks. https://ephemerajournal.org/contribution/coworking-city

Merkel, J. (2019a). Curating strangers. In R. Gill, A. C. Pratt, & T. E. Virani (Eds.), *Creative hubs in question. Dynamics of virtual work* (pp. 51–68). Palgrave Macmillan. https://doi.org/10.1007/978-3-030-10653-9_3

Merkel, J. (2019b). 'Freelance isn't free.' Co-working as a critical urban practice to cope with informality in creative labour markets. *Urban Studies, 56*(3), 526–547. https://doi.org/10.1177/0042098018782374

Mitchell, K., Marston, S. A., & Katz, C. (2003). Introduction: Life's work: An introduction, review and critique. *Antipode, 35*(3), 415–442. https://doi.org/10.1111/1467-8330.00333

Morrow, O., & Dombroski, K. (2015). Enacting post-capitalist politics through the sites and practices of social reproduction. In K. Meehan & K. Strauss (Eds.), *Precarious worlds: New geographies of social reproduction*. University of Georgia Press.

Nakano, D., Shiach, M., Koria, M., Vasques, R., Gomes dos Santos, E., & Virani, T. E. (2020). Coworking spaces in urban settings: Prospective roles? *Geoforum, 115*, 135–137. https://doi.org/10.1016/j.geoforum.2020.04.014

Orel, M., & Alonso Almeida, D. M. (2019). The ambience of collaboration in coworking environments. *Journal of Corporate Real Estate, 21*(4), 273–289. https://doi.org/10.1108/JCRE-12-2018-0050

Ostrom, E. (1990). *Governing the commons: The evolution of institutions for collective action*. Cambridge University Press. https://doi.org/10.1017/CBO9780511807763

Resch, B., Hoyer, P., & Steyaert, C. (2021). Affective control in new collaborative work: Communal fantasies of purpose, growth and belonging. *Organization Studies, 42*(5), 787–809. https://doi.org/10.1177/0170840620941616

Richardson, L. (2016). Sharing knowledge: Performing co-production in collaborative artistic work. *Environment and Planning A: Economy and Space, 48*(11), 2256–2271. https://doi.org/10.1177/0308518X16653963

Sandoval, M. (2016). Fighting precarity with cooperation? Worker co-operatives in the cultural sector. *New Formations, 88*(88), 51–68. https://doi.org/10.3898/NEWF.88.04.2016

Schmidt, S., & Brinks, V. (2017). Open creative labs: Spatial settings at the intersection of communities and organisations. *Creativity and Innovation Management, 26*(3), 291–299. https://doi.org/10.1111/caim.12220

Smith, T. S. J. (2020). 'Stand back and watch us': Post-capitalist practices in the maker movement. *Environment and Planning A: Economy and Space, 52*(3), 593–610. https://doi.org/10.1177/0308518X19882731

Spinuzzi, C. (2012). Working alone together: Coworking as emergent collaborative activity. *Journal of Business and Technical Communication, 26*(4), 399–441. https://doi.org/10.1177/1050651912444070

Vidaillet, B., & Bousalham, Y. (2020). Coworking spaces as places where economic diversity can be articulated: Towards a theory of syntopia. *Organisation, 27*(1), 60–87. https://doi.org/10.1177/1350508418794003

Waters-Lynch, J., & Duff, C. (2021). The affective commons of coworking. *Human Relations, 74*(3), 383–404. https://doi.org/10.1177/0018726719894633

Part III
Alternative Modes of Development Through Coworking and Policy Responses

The Abundant Crops of Subjectivisation and the Difficult Arts of Institutionalisation: Building Transformative Collaborative Spaces of Urban Labour in the Context of the Mares de Madrid Initiative

Alessandro Coppola

Introduction: Questioning the Transformative and Institutionalisation Outcomes of Collaborative Urban Labour Spaces

The reinvention of place and spatial relations has been quintessential to the development of contemporary social innovation practices, new urban movements and neo-municipalist discourses. New coworking spaces and spaces of labour at large have played an important role in the shaping of their imaginaries and projects. In this contribution, we will present and discuss the coworking spaces and broader labour-based collaborative processes promoted by the "Mares de Madrid" (MdM, hereafter) initiative started by the city of Madrid under the tenure of the progressive, neo-municipalist coalition of Mayor Manuela Carmena between 2015 and 2019. Funded by the EU Urban Innovative Actions (UIA) programme, MdM built on the legacy of a variety of grassroots, collaborative practices that developed in the city as alternatives to austerity policies put in place by the state in response to the 2008 great recession. MdM aimed to create neighbourhood-based and labour-intensive economies focusing on mobility, recycling, food, energy and care through the start-up of social and solidarity economy (SSE hereafter) entities to be incubated in new collaborative workspaces reclaimed from abandoned buildings. We will examine this case moving from an institutional perspective, looking at how practices such as the ones conveyed by Mares MdM can be activated, produce changes within the local system and eventually consolidate in the long term through processes of *institutionalisation*.

A. Coppola (✉)
Department of Architecture and Urban Studies, Politecnico di Milano, Milan, Italy
e-mail: alessandro.coppola@polimi.it

J. Merkel et al. (eds.), *Coworking Spaces*, https://doi.org/10.1007/978-3-031-42268-3_10

MdM positioned itself at the crossroads of a variety of contemporary processes and forces, allowing us to look at its trajectory across a variety of critical analytical dimensions while contributing to a variety of debates. In the 2010s' Madrid's context, new social and urban movements led among certain social groups to new processes of subjectivisation, to be understood as "existential forms of resistance and assertion of subjectivity (...) and an individual search for alternatives to counter the effects of systemic forces" also through the creation of new "living constructs", i.e. "activities that reinforce the construction of (...) personal identities while, at the same time, sharing in the creation of a collective identity" (Farro, 2014, p. 23). Neo-municipalist political discourses and projects have offered these processes a credible trajectory of expansion, if not hegemony, and a platform to experiment with and nurture such new living "constructs". Transformative social innovation frameworks have supplied a wide variety of epistemic communities, elements of knowledge, tools and ways of doing essential to their actual codification and implementation. Finally, coworking practices have offered a term of reference for designing new communal spaces and articulating spatial and scalar arrangements from the single building to the city and beyond. These processes and forces are not intended as a causal, diachronic chain but rather as a complex, multi-layered process in which every single element has provided multiple, mutual, retroactive feedback to initiatives such as MdM. Resulting from these processes and forces, this initiative's trajectory can be observed at various interconnected scales and axes.

On the one hand, we can look at how it changed entrenched balances of power by redistributing it between different social groups and between established formal institutions and other actors. On the other, we can look at how it advanced a new spatial imaginary able to generate rescaling from the very micro-level – that of the single facilities it created – to the neighbourhood and urban/metropolitan levels. The varying degree of MdM success in assuming a transformative position across scales and along these axes signals the level of effective institutionalization it achieved. By institutionalization, we intend the change of patterns of interaction based on the transformation of values, rules, procedures and actors involving the political and policy realms as well as civil society (Sánchez-Hernández & Glückler, 2019). Naturally, the institutionalization of a practice depends on various factors that deal with the characteristics of the governance and regulative system and the social and urban structure. The contribution is organised as follows: first, we review relevant literature on social innovation, neo-municipalism and coworking spaces; second, we present MdM's context and main components; third, we look at workings and spatialities of the MdM's facilities; fourth, we discuss the materials presented; and fifth, we draw some conclusions and formulate some further research questions.

Common Questions for Different Debates: The Quest for Transformation and Institutionalisation in Social Innovation, Neo-municipalism and Collaborative Working Projects

To be considered "transformative" social innovation (SI, hereafter), practices have to "challenge(s), alter(s) or replace(s) dominant institutions in the social context producing an irreversible, persistent adjustment in societal values, outlooks and behaviours" (Avelino et.al., 2019, p. 196). Westley and Antadze (2010) define SI as "a complex process of introducing new products, processes or programs that profoundly change the basic routines, resource and authority flows, or beliefs of the social system in which the innovation occurs" (p. 2). Transformative SI practices are the ones able to rearticulate the economic realms based on new social and territorial logics and needs (Brody et al. 2021), putting into question inherited development patterns in ways that can be both radical and incremental (Bauler et.al.2019). They also empower marginalised social groups and challenge power asymmetries while introducing institutional changes in governance systems (Smith, 2017). A "multi-actor perspective" (Avelino et al., 2019) in the analysis of SI practices has illuminated the emergence of instances of so-called bottom-linked governance (Medina-García et al., 2021), to be understood as "new forms of democratic governance collaboratively built between initiatives and activists, their scalarly dynamic networks and state institutions and agencies" (Moulaert et al., 2019, p. 4). These arrangements can combine a variety of forms of democratic innovations, from co-production models of design and supply of services to deliberative arenas. As they have been characterized by a quest for new sources of local political legitimisation, neo-municipalist political projects and discourses have commonly been platforms for the experimentation and spread of such innovations. They have done so in inherently spatial ways, rethinking scalar arrangements to enhance local participation and deploying new ideas regarding the organisation of production and social reproduction in the urban space. New municipalist experiments have also prefigurated "alternative regionalisms" attempting to challenge state-regulated capitalism and incubate alternative economic spaces (Thompson, 2021). The neo-municipalist wave in Spain articulated claims on the overall functioning of the political and economic system at the national and European levels with expectations of a change of entrenched urban development models and the ways in which production and social reproduction are organised in cities. Both SI, grassroots practices developed in Spanish cities in the context of the so-called 2011 indignados/15M movement and the following neo-municipalist political projects have been characterized by the protagonism of a young, highly precarized and skilled workforce – what we can define as the urban cognitariat – whose prospects have been put even more into question by the impacts of austerity on employment, welfare and education. Such forms of collective action became the source of relevant processes of subjectivisation for this social group (Farro, 2014), while its mobilisation in neo-municipalist political projects has also represented a way to channel the largely untapped reservoir of collective intelligence they represent into forms of policy co-production (Coppola, 2018a, b, c). SI initiatives and neo-municipalist projects have also addressed issues related to the ways and spatialities of urban labour experimenting with new types of coworking practices. Traditionally, coworking spaces have been understood as the sharing of the workspace between individuals working independently, who are heterogeneous by occupation, sector and organizational status and affiliation and develop mutual relationships either spontaneously (Gandini, 2015; Orel & Dvouletý, 2020) or as the outcome of specific activities and tools (Parrino in Gandini & Cossu, 2021). In this regard, their nature of platforms aimed not only at overcoming the solitude of post-Fordist, creative and independent urban workers but also at responding to "individualised worker's demands for autonomy, recognition and social belonging" (Merkel, 2019, p. 529). Coworking spaces acted as informal mechanisms for the development and circulation of tacit knowledge embedded in knowledge-intensive working practices (Moriset, 2013) in the context of de-structuring labour relations. As "urban socio-material infrastructures" that "organise interaction within them but also with one another, enabling networks of communication", they can also provide a platform for economic, political and social action" (Merkel, 2015, p. 123). How these essential qualities are theorized and performed surely also depends on the political discourses and narratives of the "community" within which coworking spaces are embedded and on the local, institutional and regulative contexts (Spinuzzi et al., 2019). As coworking spaces are finely "curated" spaces and practices, the ways in which the intentional creation of interconnections between people, ideas, objects and places is made are critical. Building on the three strata definition of the creative city of Cohendet et al. (2010) made of "underground" (skilled individuals), "middle-ground" (epistemic communities and communities of practice) and "upper ground" (institutions and organizations), Gandini & Cossu (2021, p.435) locate coworking spaces in the middle ground, also noticing how their role of mediation include relationships with neighbourhoods. The existence and nature of this relationship have become of particular importance for discriminating between coworking practices compatible with forms of "collaborative individualism", commonly mobilised by real-estate promoters and corporations, and "resilient coworking practices" attempting to "blending entrepreneurial logics with forms of political and social activism" (Gandini & Cossu, 2021). The latter work proactively to maintain a closer relationship between the space and its local context, producing outcomes that do not reproduce the individualised ethos the neo-corporate model engenders while promoting a community discourse based on social impact (Gandini & Cossu, 2021). Location is vital in this regard. Coworking spaces' location patterns often resemble those of service industries in urban areas, with so-called creative clusters representing a preferential location (Di Vita & Mariotti, 2017). As a policy tool, coworking spaces have been deployed to create variably scaled economic clusters or to achieve wider societal goals such as the "urban regeneration" of peripheries and a better linkage of their lower-income residents to economic opportunities. When comparative costs drive locational choices outside of already developed areas, coworking spaces are deemed to facilitate gentrification, supporting the symbolical repositioning of less central areas. At the same time, locating coworking spaces in outright peripheral areas incurs a risk of insulation from the local context. These different streams of literature provide relevant elements for the analysis of MdM's context and development that we will present in the next section.

Mares de Madrid in Its Context: From New Social Movements to Neo-municipalist Politics

In 2015, ending a decades-long political and electoral hegemony of the neoliberal, conservative Partido Popular (PP), a progressive political figure – Manuela Carmena – was elected mayor of Madrid. Her victory was the outcome of wider political processes involving the entire country. Starting in 2014, indignados/15M movement's activists formed new progressive urban electoral coalitions in connection with the national rise of *Podemos*, a new "left populist" political force. This convergence was presented as the embodiment of a new political practice of "confluence" (confluencia) between parties and organisations, ordinary inhabitants and grassroots groups. This was the outcome of a process of "overflow" (desborde), meaning a new, unpredictable logic of collective action leading to innovative self-organised forms of protest – such actions in defence of public education or healthcare – and innovative institutional initiatives (Rubio-Pueyo, 2017). The creation of *Ganemos Madrid* and its confluence in a larger electoral cartel – *Ahora Madrid* (AM, hereafter) – was one of the main events in this process of political restructuring deeply rooted in previous patterns of social mobilisation. AM's programme had some key, defining characters. First, moving away from a conventional framing of administrative and policy realms, the coalition advanced new thematic priorities underlining the mutual connections between various issues such as inequalities, unemployment, the environment, neighbourhoods, gender, and citizens' participation. Second, besides using primaries to select candidates, AM proposed a participative approach to draft its platform integrating neighbourhood assemblies and online participation tools by following a cooperative open-source logic (Rubio-Pueyo, 2017). Third, the transformation of urban governance through neighbourhood decentralization, increased accountability and direct participation was paramount in the coalition's discourse. These elements resounded with the key elements of the processes of subjectivisation that wide components of the "cognitariat" (Rubio-Pueyo, 2017) experienced in the context of the indignados/15M movement. AM's success proved the maturity and relevance of these subjectivisation processes and the possibility of linking the interests and self-representation of the cognitariat with the ones of the more traditional working classes and the peripheral neighbourhoods in which these lived. Once in power, the new coalition promoted various grand strategies embodying a decisive change in local political discourse. A specific commissioner for neighbourhood's participation was appointed, the local districts' role was strengthened with the formation of local, thematic working groups channeling policy proposals towards the city, and a "territorial cohesion fund" ("Fundo de reequilibrio territorial") aimed at increasing the allocation of city funds to more disadvantaged neighbourhoods was launched. Also, a participatory budgeting initiative, the Decide Madrid initiative, was implemented (Feenstra & Tormey, 2021). The city launched a new SSE strategy aimed at supporting new cooperatives and social enterprises while pursuing social and systemic innovation through the experimentation of co-production of municipal policies. The strategy pursued the creation of SSE territorial, neighbourhood-based ecosystems, and a comprehensive support system for creating new businesses (Arampatzi, 2022). Following a wider trend of politicization of care at the crossroads of recent feminist movements and grassroots social practices grown during the great recession, the new administration launched a "City of Care' strategy" – La Ciudad de los cuidados – as one more cross-sectoral policy aimed at influencing health, education, culture, and environment policies (Martinez & Gasperi, 2021). The plan "sought to implement pilot projects and prototypes to be tested and improved at the neighbourhood level and from a community perspective" (Martinez & Gasperi, 2021, p. 23), favouring the collaborations between care providers and the care demand and the innovative use of public facilities. Governance innovation and hybridisation were one of the goals, as the social-public partnership model it promoted intended to be an alternative to new public management style public-private partnerships that, also in care, had been quintessential to decades of neoliberal governance in Madrid. In many ways, the MdM initiative acted as an ambitious pilot project engaging with the multiple trenches from which the neo-municipalist administration was attempting to change local equilibria while mobilising social groups and activists' network that had been critical to its electoral success. In this sense, it appeared not to be an insulated, opportunistic project aimed at simply capturing EU funding but rather the most comprehensive, integrated attempt to embody in one single initiative a variety of themes at the centre of the neo-municipalist discourse: a condition that fuelled its potential but that also made it dependent on the success and institutionalization of other key policy and regulative changes.

Skilfully Planning Transformation: Governance, Frameworks and Tools of the Mares de Madrid Initiative

MdM pursued a variety of objectives at once: the prototyping and co-designing of new functions for disused public buildings, the fostering of territorial economic innovation processes and ecosystems based on the adoption of SSE models and the creation of new and better-quality employment opportunities in working-class, high unemployment neighbourhoods (Mares de Madrid, 2019). The centerpiece of the initiative was the creation of new facilities called "Mar" located in the neighbourhoods of Centro, Villaverde, Vallecas and Vicálvaro, each of them characterized by a certain degree of concentrated social disadvantage and potential specialisation in one of the sectors of mobility, food, recycling, energy and care. The Mar was to become the most advanced prototype of the multi-purpose territorial facilities envisioned in the mentioned SEE strategy, including spaces for individual and group work, collective and neighbourhood-based activities of different sorts and production . In line with UIA requirements, MdM moved from a partnership stretching across scales – from the city to the region – and across different types of actors, state entities, civil society, professionals and companies, with at the core a strategic relationship between the city administration and a set of entities with variably intense ties with indignados/15M movement and grassroots practices, including the cooperatives *Dinamia*, *Tangente*, the architecture firm *Sic Arquitectura y Urbanismo* and the NGO *Vivero de Iniciativas Ciudadanas* (VIC). MdM governance model embodied that new way to conceive governing the city along the lines of a new so-called "public-social" collaboration, encompassing on virtually equal footing the city administration on one side and civil society and SSE actors on the other. The initiative was governed by a board including all coalition participants – the Mares coordination – with the most relevant tasks being carried out by the mentioned cooperatives and associations and city officials focusing instead on the more regulation-burdened tasks of procuring the works for the renovation of the four buildings. The collaborative nature of the endeavour was also visible in the key role assigned to SSE partners in the management of the new facilities. This approach drew its legitimacy not exclusively in the political discourse of the new city administration but also in MdM's very high degree of organisational complexity and innovation that, as such, implied a high degree of flexibility, learning-by-doing and cultural capital hard to find in austerity weakened, ageing public administrations. MdM was organized around a set of intermingled processes aimed at building evidence and capabilities among the actors involved toward creating new SEE businesses able to form highly specialised, innovative and inclusive ecosystems. All these processes were centred around specific, highly theorised devices mostly oriented at turning forms of tacit knowledge into strategic resources within the framework of a new, intensively communicative and relational *milieu* participated by residents, entrepreneurs, activists and experts embedded in both local and supra-local networks. Such devices moved from a peculiar understanding of local development policy as a process in which policymakers had to act as activists engaged in constantly travelling between different scales and the demand and supply sides to construct a new economic sector. This represented a wholesome approach with ambitious and ramifying implications in policy areas as diverse as training, employment, neighbourhood services and welfare. Among these tools were the so-called skills laboratories ("laboratorio de competencias"), value chains ("cadenas de valor") and practice-oriented learning communities ("comunidades de aprendizaje orientadas a la practica"). The "Laboratorios" were designed by Vivero de Iniciativas Ciudadanas and Dinamia based on previous experiences of collaborative mapping focusing on evictions and self-help practices (Gonick, 2016; Annunziata & Lees, 2016). The first step of the *laboratorios* was the co-creation of four "cartografias ciudadanas" (citizens' cartographies), one for each neighbourhood, based on residents' and local actors' inputs on the formal and informal grassroots initiatives that could be associated with Mares' five thematic cores. Later, the laboratories moved on to mapping existing , unaddressed social needs and the inhabitants' skills. Also, the so-called Cadenas de valor ("value chains") were visual representations of the system of relations and processes connecting economic actors participating in the same market from the raw materials' provision throughout the final consumption of a specific product. In reference to "circular economy" conceptualisations, Mares' "Cadenas de valor" also included so-called cascades, meaning how a product, once consumed, could make its return to the origin of the entire chain and suggestions regarding the development of "do-it-yourself" (DIY) and collaborative economies by allowing the conflating of different rings of the chain (as an example: repairing your bike, growing your food). The "Cadenas" were studied both at the sectorial level – mobility, food, energy, recycling and care – , looking at how articulated and complete were the productive networks in the four neighbourhoods. Representing both the current economy and the potential for transformation, the design of each "value chain" was organized around a matrix of relevant "badges", including products and services already present in the territory, the current role of the social economy, the services promoted by the public administration and products and services produced off the market. The "Cadenas" were built through workshops involving promoters of new SSE projects, local actors and established actors active in each productive chain. Coming to the latter tool, by "Comunidades de aprendizaje orientadas a la practica" ("practice-oriented communities of learning", CAP), the Mares initiative understood "groups of people who share a concern or passion about something they do and learn how to do it better as they interact" (Mares de Madrid, 2019). Based on Etienne Wenger's theorisations (Wenger, 2004), CAPs were frameworks for research, generation, acquisition and deepening of knowledge oriented towards structuring the SSE initiatives MdM intended to support. Knowledge was to be produced from the interaction between community members but also relying, when needed, on external inputs. In so doing, CAPs attempted to do what in workforce development policies was done through traditional formal education and training. CAPs had to be open-ended, and their domain could be related to a topic (e.g. childhood), to a know-how (e.g. how to manage teams). Their participants could range from individuals starting a project to established specialized actors or activists interested in building knowledge, not necessarily from a perspective of immediate economic use. All these processes were supposed to mutually contribute to the social production of new localised economic ecosystems made of an evolving social demand, forms of everyday knowledge as well as specialised expertise. The "laboratories" were supposed to generate new possible economic projects based on mapping local needs and skills. "Cadenas" were supposed to indicate niches for developing new economic projects and opportunities for collaboration and interdependencies between the value chain of a specific product with those relative to the other sectors involved in MdM's operations. Finally, the CAPs were supposed to allow economic projects to tap into a large reservoir of expert knowledge left at the margin by an economic system that seldom involved the most educated, both formally and even more informally, sections of the population. All these processes were initiated, at times completed, while the city was working on opening the four facilities.

Negotiating Space, Reshaping Scale: The *Mar* as Places for the Social Production of the Urban Economy

The opening of the four *Mar* spaces was a key objective of the initiative, as the materiality of the spaces was also set to embody the shaping of a relationship between all these sophisticated lines of work with real places – especially the neighbourhoods – and users. However, since the beginning, the choice to place the four new spaces in decommissioned public buildings proved problematic as rising costs, archaeological findings and other public procurement issues impeded the opening of the Mar of recycling in the neighbourhood of Vicalvaro and the Mar of Mobility in Vallecas. For the latter two, the city rushed to set up temporary solutions in the same districts, while the Mar of Energy in Centro and the Mar of Food in Villaverde were instead opened in 2019. Also, provisional spaces of a more productive nature – a makerspace connected to the Mar of Energy and a logistics space connected to the Mar of Mobility – were set up by renting private spaces. Since their gestation through a co-creational process managed by VIC, the Mar were supposed to be hybrid spaces combining functions, uses and social groups rarely cohabiting under one roof: a space of work and collaboration for the incubated economic projects, of early access for people interested in the SSE and in the five economic sectors, of learning and training, of organising and animating the relationship between the initiative, the projects and the community. Exclusively for reasons of conciseness, we will only focus on the Mar of Energy and the Mar of Food while setting aside the the other mentioned spaces.

The project of the Mar of Energy reclaimed a community centre built in the 1980s, creating a flexible space organised around three main functional areas. In the coworking area, participants in the specific economic projects shared an open space with no fixed places while having the opportunity to use reserved rooms for internal meetings, meetings with clients and with other projects or with the MdM support services. Social events, from training sessions to workshops promoted by local groups, took place instead in a collective space. Economic projects had to be assessed by the Mares coordination to be incubated within the Mar. The incubation contract lasted one renewable year to ensure its transitional nature and the ability to accommodate new projects continuously. Projects had to be coherent with the thematic concentration, even if cross-sectorialism and experimentalism were highly valued. Incubated projects involved highly skilled, mostly young individuals and included initiatives like La Corriente, a renewable energy cooperative; Solencoop, a cooperative specialized in the setting up of solar energy systems; and Socaire, an association focusing on energy with an emphasis on fighting energy poverty. The management of the space was produced in a rather flexible way through the combination of formal and informal mechanisms that, besides their consistency with the goals of the initiative, were based on the aim of "building a sense of ownership and belonging among the regular users of the space" (Coppola, 2019). Each Mar had a coordinator appointed by the Mares coordination before the start of the initiative and was normally part of one of the promoting organizations. A Mar coordination committee met monthly and was participated by the coordinator, the members of the different working areas of MdM and the participants in the incubated economic projects. The committee shared an agenda for the use of the space while taking fundamental decisions about its planning and daily workings. At the same time, self-organisation contributed to the facility's management, with users proposing initiatives and resolving daily issues – including general upkeep, the use of the kitchen, etc. – relying on a limited budget provided by the city.

The Mar of Food in Villaverde responded to a similar spatial design logic. However, it also included specialised facilities, such as a 150 square meters kitchen called the Gastrolab and storage space. Besides, there were coworking spaces like the ones of the Mar of Energy, a shared meeting space and a terrace open to users' gatherings. In this case, too, the Mar coordination approved the incubation of the projects and, therefore, their access to the Gastrolab based on some essential criteria: the project's maturity, social impact, and ability to strengthen the Mar's sectorial specialisation. Projects were grouped into three main families: the *pre-viveristas*, i.e. informal activities with a minimal level of planning; the *viveristas*, i.e. formal activities close to marketisation; and finally, initiatives aimed at the involvement of the neighbourhood. The Gastrolab had its own coordinator, who was also responsible for the training of economic projects' members in the management procedures, health and safety rules and the sustainability of the productive processes. The overall goal of the Mar of Food was to support new SSE economic projects all along the entire food chain with an emphasis on local, sustainable, fair and healthy food. At times, this implied – as also happened in other sectors – to push individual projects to become collective projects and promote collaborations between similar projects that could, in this way, increase their scalability. These aggregation processes could happen before and during the access to the Mar, being both a precondition and an outcome of using the space. Furthermore, the Mar of Food also facilitated formalising informal economic practices. Projects supported by the Mar of Food included La Osa, a cooperative supermarket; Hortesana, an agricultural cooperative; and Las Hermanas Budares, a food production cooperative formed by women with a Venezuelan background.

In both cases, the relationship with the surrounding community was the object of great attention and reflection by the coordinators and the project coordination . The relation between the Mar as a "community place" open to all and a specialised place for the development and clustering of economic projects and, more in general, the tension between a territorial and a sectorial understanding and operationalisation of the initiative were one of its most complex challenges and, as such, had to be continuously negotiated. The overall aim of the "Agitacion economica" line of work was to orchestrate the active involvement of an extensive series of actors across territories and sectors in support of the economic initiatives. The first step was the mapping and outreach to key actors, including district governments and local employment boards. In the case of the former, the focus was the establishment of so-called "mesas public-social" – collaborative schemes between the districts and actors of the SSE – with the identification of potential public procurement innovations. The rationale of the four neighbourhood strategies designed by the team was, in general, the development of a "social market" at the local level: local actors could support the creation of local demand for the services of the incubated economic projects while acting as a gateway to a certain social group – i.e. migrants, the elderly, unemployed – or for sensibilisation activities regarding the SSE. Neighborhood conditions, in this regard, varied greatly.

For the Mar of Energy, the main risk was, in fact, "to die of too much success" (Coppola, 2019) as its central location and thepresence of the young cognitariat and "consumption aware" middle classes exercised a positive pressure on the new centre, making it a good starting point for the set-up of many of the economic projects incubated. Outreach to the population was one of the key activities of the Mar by building relationships with trade associations and local groups and participating in local fora and organizations, employment boards and schools, as well as building a market for the products and services offered by its economic projects. By the end of the initiative, the Mar of Energy had become a pivotal location for people interested in energy transitions by hosting initiatives, facilitating exchange and collaboration between economic projects and activists and animating outreach initiatives and learning based on the deployment of the tools we mentioned. In Villaverde, the strategic location of the Mar along its main street , its design and aesthetics made it an evident presence, although, differently from the Mar of Energy, the social composition of the area did not make new links based on consumer awareness as easily as in Centro. Open-house days were organised to inform residents about the Mar's activities to overcome the problem.

By its closure in 2019, Mares de Madrid had opened two long-term public facilities and had involved 559 economic projects, supported 332 in a tailored manner, incubated 91 and directly contributed to the foundation of 48 new businesses (Mares de Madrid, 2019). The elections taking place the same year saw the return to power of the PP with, this time, critical support of the recently formed far-right party of Vox. This political change happened at a critical juncture of MdM's implementation as, with the ending of EU funding, its consolidation had to be decided. Its fate widely depended on the positioning of the new conservative administration with the broader change in political discourse promoted by Carmena's administration through that variety of grand, although highly narrative, strategies we mentioned. Although some financial support for SSE enterprises continued, the new administration sidelined the implementation of the SSE strategy: the planned "Observatory of the Social and Solidarity Economy" was not put in place, and the offices for the support of the SSE, the "Oficinas de la Economia social", were closed (Coppola, 2020). In this context, as the Carmena administration did not manage to launch a new procurement process before the elections, MdM managing entities found themselves no longer in charge of the Mar spaces that all lost their coordinators and staff. Only the Mar of Energy and Food – although not the Gastrolab – stayed open to a limited number of projects previously incubated but with no support and the insertion of other non-SSE projects supported by Madrid International, a mainstream start-up incubator.

In the longer term, the city included the Mar in the framework of a more comprehensive policy aimed at creating more traditional "innovation ecosystems" while erasing their sectoral focus and limiting the quota for SSE initiatives to a maximum of 25 percent. The management of these spaces was decided by procurement processes open to all kinds of companies in the perspective of a more formalised, top-down governance, including universities and business actors. Among the activities required within the procurement call were also devices introduced by Mares. However, there was a change in focus from establishing connections among incubated initiatives and the neighbourhoods towards establishing connections with larger companies or investors (Coppola, 2020).

Discussion: A (Too) Great Transformative Ambition and a Failed Institutionalization

Starting with the first axis of our analysis and looking at it against the legacy of Madrid's long-term neoliberal governance and regulatory model, we can say that MdM has been a relevant source of innovation in policymaking and governance models and of empowerment of specific social groups. MdM's management was mostly directed by a relatively younger, high-in-cultural capital urban cognitariat that was long marginalised in the city's governance and regulative system before the advent of the neo-municipalist administration. Although on several accounts, this social group was in a more privileged position than the working classes that the initiative intended also to serve, it was exposed to high unemployment and precarity further dramatised by the effects of austerity . Other social groups were involved and empowered as well – a variety of economic projects included individuals with backgrounds with ties with low-skilled immigration and the working classes (see the case of the Las Tres Calles cooperative in the area of care in Coppola, 2018a, b, c), but they were certainly in a lesser central position in MdM's operations. This outcome is largely explained by the previous subjectivisation processes that involved the urban cognitariat. These represented a pre-condition for MdM's design and legitimisation as a policy tool while arguably becoming the most critical resource throughout its implementation. The overall initiative and the Mar spaces were highly selective and curated environments that proved able to accommodate the demands and imaginaries of new living environments this social group brought (Farro, 2014) while managing to advance a clear transformative reading of the role of coworking spaces and collaborative spaces of labour. By selecting projects based on a set of prerequisites – the first being collective SSE economy projects – and pushing them to scale up through collaborations, merging and articulating with local needs and demands, MdM's management framed coworking spaces as spaces aiming not solely at improving working and living conditions for the cognitariat, but more broadly at the construction of alternative economies through the design and activation of a variety of complex organisational processes. Accordingly, the MdM's management model combined relatively top-down dimensions leveraging the new innovative managerial culture circulating in SI-related epistemic communities with more collaborative empowerment dynamics leveraging on subjectivisation processes that pre-existed MdM but were also further empowered in its context.

Moving to the second axis of our analysis, the spatial imaginary advanced by MdM and embodied in the opening of the Mar themselves was bold and transformational in many ways, openly framing and addressing processes of peripheralisation and polarisation shaping the urban and metropolitan context. On the one hand, it advocated for the location of sophisticated urban functions in peripheral neighbourhoods in a traditionally spatially redistributive manner. On the other, in more innovative ways , it advocated for a different understanding and operationalisation of space in the social construction of the urban economy. It did so by pursuing the establishment of neighbourhood-scale markets and systems of production. On one level, MdM was a relocalisation project – promoting a downward rescaling of certain economies – while on the other, it attempted a new form of upward rescaling by developing ties between actors based in Madrid s and other local and supra-local actors to create new SSE entities able to enter in the city's procurement systems and disrupt entrenched equilibria and incumbent positions (see the case of collaboration set-up with Basque cooperative actors by the Mar of Recycling in Coppola, 2017). The challenges that MdM encountered in these rescaling exercises also point to the mentioned critical issue of the relationship between coworking spaces and neighbourhoods. The attempt of the initiative to create a framework for the circulation, blending and mobilization of the tacit knowledge embedded in different social worlds, the management entities and the new economic projects on the one hand and the local populace on the other has proved difficult. The mapping of local inhabitants' skills seldom led to economic projects, with most of the latter developing out of urban level, higher skilled patterns of engagement. This varied across sectors and spaces – as mentioned, initiatives in care and food had a more working-class and, at times, neighbourhood-based rooting – but overall, the results in this regard were limited. The initiative had a very structured, highly designed approach to this matter, and likely the long-term routinisation of many of the related Mares' innovative practices would have brought more results in the medium term. The deployment of key resources coming from SI managerial culture in such a short time was not enough to fill a gap in subjectivisation that characterised working-class groups and the neighbourhoods in most cases. The everyday functioning of the spaces was essential in this regard as it allowed the reproduction of processes of subjectivation that were linked to the creation of new collectivities around the management of the Mar itself and the creation of new networked economic projects. However, this process did not take place at the scale of the neighbourhood in the measure initially imagined. Rather, it happened mostly at that of the epistemic "middleground" that, as mentioned, has been considered particularly relevant by the literature on creative economies and coworking. These developments, which are largely dependent on the wider social and urban structure in which the initiative took place, are to be put in relation to broader circumstances if we want to understand why the MdM's institutionalization at the political and policy level failed.

MdM was the outcome of a peculiar, transcalar convergence of circumstances: the eruption of new social movements at the national level, the spread of SI practices at the local level, a successive change in political leadership at the city level – the neo-municipalist turn – and the availability of opportunities of innovation for cities at the EU level. This powerful, temporary convergence allowed deep, intense experimentation that has also involved the reframing of public conceptualisations of urban services, the labour it implies, the spaces in which it can be nurtured and the role that institutions play in their structuring. However, especially in Madrid, the neo-municipalist platform had to be implemented under significant institutional constraints that included unknown (to new leaders) and slow bureaucratic procedures, legal limitations, misalignments with higher levels of government and austerity measures that impeded spending and recruitment (Mota Consejero & Janoschka, 2023; Blanco et al., 2020). Despite such constraints, MdM was implemented in a very experimental manner by incrementally defining a set of highly codified practices and procedures that came to represent a parallel formality to that of entrenched city procedures and routines (that had to be recognised and frequently attended). However, the political leadership was not able to give way to a variety of deeper changes in regulation that would have improved chances of institutionalization even under conditions of political change: in particular, it failed to launch a procurement process for the longer-term management of the Mar spaces based on the initiative's practices that would have institutionalised MdM's innovative governance model; to enact a substantial review of procurement policies that would have opened up market opportunities to the SSE economic projects incubated by MdM while embedding them in city procedures; and to effectively execute its decentralisation and participation policy that would have strengthened the possibility of strong integration between the Mar and the neighbourhood level (Coppola, 2018a, b, c). These failures signal the inability of the municipalist leadership to move beyond a phase in which political intervention was organized around the coupling of highly narrative interventions (the many overarching strategies presented by the administration) with the activation of temporary, louse spaces of policy experimentation (the UIA initiative) in favour of a more stable, regulation-intensive institutionalization of change. The reasons for the lack of this ability may be manifold – lack of political resources, too high regulative barriers, lack of support from a variety of actors and social groups – but they seem, however, to have played an important role in the failed political and policy institutionalization of the practices advanced by MdM. Institutionalization happened, but it was mostly at the economic level, with the lasting of several of the economic projects incubated by the initiative. The spaces in which they were incubated – that represented a recognition and institutionalization of the "living environments" brought upon by the cognitariat in the context of their previous processes of subjectivization – were instead largely normalised, coming to represent a case of failed long-term institutionalisation.

Subjectivization, Scale and the Forms of Institutionalization: Three Open Questions for Further Research on Collaborative Urban Labour Spaces and Urban Change

Progressive political strategies and their relationship with inherited governance and regulative environments, subjectivisation processes of certain social groups and their relation to broader class and social structures, resources circulating in epistemic communities and their availability to grassroots practices, inherited urban social geography and their ability to support rescaling processes: all these elements appear to be relevant in the making of local assemblages of transformative practices such as MdM and in defining their chances of institutionalization. The case of MdM proves that, unsurprisingly, such assemblages are widely mediated by local conditions and that some general questions related to the transformative potential of coworking practices and the like have to lead to profoundly context-sensitive analysis. In conclusion, we suggest three main potential research avenues that arise for the analysis of this case and that we consider relevant for the variety of debates we have engaged with.

The first issue revolves around the complex relationships between these practices and the local scale. Focusing on neighbourhoods as spaces not just for planning and regulating social *reproduction* but also for social *production* opens a range of tensions that are of great importance. If the issue of the impact that such spaces can have on the valorisation of peripheral areas has become a common concern in the literature, that of the complex relationship between policies intending to cluster certain economic activities characterised by a high degree of innovation – and more specifically social innovation – and the actual social composition of these neighbourhoods keeps being largely unaddressed. Spatial proximity does not necessarily lead to social proximity, as class identity and belonging and sectoral specialisation dynamics are principles of social organisation in cities far more important than simple spatial proximity. Unless a more robust, long-term institutional infrastructure striving for spatial equality is put in place in neighbourhoods, the creation of cross-class urban alliances and living environments between the cognitariat and the working classes will be hard to achieve through innovative, transformative practices. This is a relevant question on which the overall collective desirability of their location in peripheral areas depends.

The second, deeply related issue revolves around the role of subjectivisation processes. As the production and negotiation of social identities become increasingly central in the governing of the city – and more in particular in urban regeneration and innovation initiatives (Coppola & Lucciarini, 2023) – the uneven involvement of social groups in these processes becomes a key factor for the analysis of the trajectory and outcomes of transformative practices. As we have seen in the case of MdM, subjectivisation processes are critical both as a precondition of these practices and as a continuous, self-feeding outcome of them. However, if the transformative nature of a practice lies in its goal to include and mobilise disadvantaged social groups, this may be very hard to achieve, especially in the short term and notwithstanding the very intention of its promoters. Even if feeling responsible for broader social change, cognitariat's entrepreneurial subjects are already extremely busy enhancing complex relations among themselves to adequately support the subjectivisation of other disadvantaged social groups. In simple terms, while leveraging existing subjectivisation processes may be a recipe for success, manufacturing new ones, especially among working-class groups, may prove very difficult. This is indeed a relevant issue to further advance the research on the collective, community-like effect of specific ways to organise transformative coworkings, urban entrepreneurialism and self-employment.

The third issue revolves around the need to refine and articulate what we mean by institutionalization. The case of MdM shows how ambitious, transformative practices imply a multi-layered and multi-scalar understanding of institutionalization, looking both at the level of grand political discourses and policy decisions – that is by far the most attended by researchers – and at the more obscure but decisive level of local policymaking that encompasses things such as internal routines, offices' organization, partnership design and public procurement procedures. All these layers come with different scales – from the neighbourhood to the state – and temporalities that can act as many points of resistance to change, notwithstanding, in a certain measure, the level of political commitment. This pushes us to consider institutionalization as a complex game with multiple intermingled entries that transformative practices have to be able to navigate skilfully. At the same time, it also pushes us to not limit our sight to the higher political and policy levels of institutionalization as if, even if they do not manage to enter from the main door of policy directives and procedures, the outcomes of transformative practices can institutionalise in other realms, such as the urban economy and civil society.

References

Annunziata, S., & Lees, L. (2016). Resisting "austerity gentrification" and displacement in Southern Europe. *Sociological Research Online, 21*(3), 148–155. https://doi.org/10.5153/sro.4033

Arampatzi, A. (2022). Social solidarity economy and urban commoning in post-crisis contexts: Madrid and Athens in a comparative perspective. *Journal of Urban Affairs, 44*(10), 1375–1390. https://doi.org/10.1080/07352166.2020.1814677

Avelino, F., Wittmayer, J. M., Pel, B., Weaver, P., Dumitru, A., Haxeltine, A., & O'Riordan, T. (2019). Transformative social innovation and (dis) empowerment. *Technological Forecasting and Social Change, 145*, 195–206.

Bauler, T., Ruijsink, S., & O'Riordan, T. (2019). Transformative social innovation and (dis) empowerment. *Technological Forecasting and Social Change, 145*, 195–206. https://doi.org/10.1016/j.techfore.2017.05.002

Blanco, I., Salazar, Y., & Bianchi, I. (2020). Urban governance and political change under a radical left government: The case of Barcelona. *Journal of Urban Affairs, 42*(1), 18–38. https://doi.org/10.1016/j.eist.2020.01.009

Bródy, L. S., Coppola, A., D'Ovidio, M., & Pradel, M. (2021). Social innovation in Italian cities: a southern European perspective. *Social innovation in Italian cities: A southern European perspective, Territorio, 2021*(99), 53–55.

Cohendet, P., Grandadam, D., & Simon, L. (2010). The anatomy of the creative city. *Industry and innovation, 17*(1), 91–111.

Coppola, A. (2017). *The Mares de Madrid Journal 1, Urban innovative actions*. Retrieved October 20, 2022, from https://www.uia-initiative.eu/sites/default/files/2017-12/FINAL%20VERSION_Madrid.pdf

Coppola, A. (2018a). *The Mares de Madrid Journal 2, Urban innovative actions*. Retrieved October 20, 2022, from https://www.uia-initiative.eu/sites/default/files/2018-05/Madrid_01-598%20MARES_Alessandro%20Coppola_Journal%202_May%202018.pdf

Coppola, A. (2018b). *The Mares de Madrid Journal 3, Urban innovative actions*. Available at https://www.uia-initiative.eu/sites/default/files/2019-02/Madrid-MARES-Journal%203.pdf

Coppola, A. (2018c). *The Mares de Madrid Zoom-In, Urban innovative actions*. Retrieved October 20, 2022, from https://uia-initiative.eu/sites/default/files/2018-10/Madrid%20-%2001-598%20MARES%20-%20Alessandro%20Coppola%20-%20Zoom-in%201%20-%20Sept%202018_Final.pdf

Coppola, A. (2019). *The Mares de Madrid Journal 4, Urban innovative actions*. Retrieved October 20, 2022, from https://www.uia-initiative.eu/sites/default/files/2019-06/Madrid_MARES_Journal%204.pdf

Coppola, A. (2020). *The Mares de Madrid Journal 5, Urban innovative actions*, Retrieved October 20, 2022, from https://uia-initiative.eu/sites/default/files/2020-05/Madrid_MARES%20journal.pdf

Coppola, A., & Lucciarini, S. (2023). Eventually detached, eventually belonging. A residential narratives` based institutionalist perspective on urban regeneration and the middle classes in Milan and Marseille. *Cities, 132*, 104052. Advance online publication https://doi.org/10.1016/j.cities.2022.104052

Di Vita, S., & Mariotti, I. (2017). Coworking spaces in Milan: Location patterns and urban effects. *Journal of Urban Technology, 24*(3), 47–66. https://doi.org/10.1080/10630732.2017.1311556

Farro, A. L. (2014). A new era for collective movements: The subjectivization of collective action. Reimagining social movements: From collectives to individuals. In A. L. Farro & H. Lustiger-Thaler (Eds.), *Reimagining social movements: From collectives to individuals* (pp. 15–34). Ashgate Publishing.

Feenstra, R. A., & Tormey, S. (2021). From social mobilization to institutional politics: Reflecting on the impact of municipalism in Madrid and Barcelona. *Social Movement Studies*, 1–19. https://doi.org/10.1080/14742837.2021.1993181

Gandini, A. (2015). The rise of coworking spaces. A literature review. *Ephemera, 15*(1), 193–205.

Gandini, A., & Cossu, A. (2021). The third wave of coworking: 'Neo-corporate' model versus 'resilient' practice. *European Journal of Cultural Studies, 24*(2), 430–447. https://doi.org/10.1177/1367549419886060

Gonick, S. (2016). Indignation and inclusion: Activism, difference, and emergent urban politics in post-crash Madrid. *Environment & Planning D, 34*, 209–226. https://doi.org/10.1177/0263775815608852

Mares de Madrid. (2019). *Memoria final. 2016-2019*. Available at: https://maresmadrid.es/actualidad/memoria-final-mares/

Martinez, A. W., & Gasperi, F. D. (2021). Who cares for what? Care networks and new urban activisms in Madrid: Restating solidarity. *Culture, Practice & Europeanization, 6*(1), 15–38. https://doi.org/10.5771/2566-7742-2021-1-15

Martínez, M. A., & Wissink, B. (2022). Urban movements and municipalist governments in Spain: alliances, tensions, and achievements. Social Movement Studies, 21(5), 659–676. https://doi.org/10.1080/14742837.2021.1967121

Medina-García, C., de la Fuente, R., & Van den Broeck, P. (2021). Exploring the emergence of innovative multi-actor collaborations toward a progressive urban regime in Madrid (2015–2019). *Sustainability, 13*(1), 415–444. https://doi.org/10.3390/su13010415

Merkel, J. (2015). Coworking in the city. *Ephemera, 15*(2), 121–139.

Merkel, J. (2019). 'Freelance isn't free.'Co-working as a critical urban practice to cope with informality in creative labour markets. *Urban Studies, 56*(3), 526–547.
Mota Consejero, F., & Janoschka, M. (2023). Transforming urban democracy through social movements: The experience of Ahora Madrid. *Social Movement Studies, 22*(3), 343–360.
Moriset, B. (2013). *Building new places of the creative economy. The rise of coworking spaces.* Retrieved June 24, 2022, from https://shs.hal.science/halshs-00914075/document
Moulaert, F., MacCallum, D., Van den Broeck, P., Garcia, M., Howaldt, J., Kaletka, C., Schröder, A., & Zirngiebl, M. (2019). Bottom-linked governance and socially innovative political transformation. In *The social innovation landscape. Global perspectives. Atlas of social innovation. Second volume: A world of new practices* (pp. 62–65). Oekom Verlag.
Orel, M., & Dvouletý, O. (2020). Transformative changes and developments of the coworking model: A narrative review. In *Technological progress, inequality and entrepreneurship* (pp. 9–27). https://doi.org/10.1007/978-3-030-26245-7_2
Rubio-Pueyo, V. (2017). *Municipalism in Spain from Barcelona to Madrid, and beyond.* Rosa Luxemburg Stiftung.
Sánchez-Hernández, J. L., & Glückler, J. (2019). Alternative economic practices in Spanish cities: From grassroots movements to urban policies? *European Planning Studies, 27*(12), 2450–2469. https://doi.org/10.1080/09654313.2019.1644295
Smith, A. (2017). *Social innovation, democracy and maker spaces.* SWPS 2017-10. Retrieved June 24, 2022, from https://ssrn.com/abstract=2986245
Spinuzzi, C., Bodrožić, Z., Scaratti, G., & Ivaldi, S. (2019). "Coworking is about community": But what is "community" in coworking? *Journal of Business and Technical Communication, 33*(2), 112–140. https://doi.org/10.1177/1050651918816357
Thompson, M. (2021). What is so new about new municipalism? *Progress in Human Geography, 45*(2), 317–342. https://doi.org/10.1177/0309132520909480
Wegner, E. (2004). *Communities of practice. Learning, meaning and identity.* Cambridge University Press.
Westley, F., & Antadze, N. (2010). Making a difference: Strategies for scaling social innovation for greater impact. *The Innovation Journal: The Public Sector Innovation Journal, 15*(2), 1–19.

Public Support for New Working Spaces in Italy: The Case of Presidi di Comunità

Ilaria Mariotti, Ivana Pais, and Francesca Ciccarelli

Introduction

The advent of digital technology has contributed to the rise of various alternatives to traditional workplaces, such as coworking spaces (CSs), where especially self-employed and small businesses can take advantage of opportunities for social and professional interaction, reducing the risks of social isolation associated with home-working and building personal and professional relationships (Akhavan, 2021; Merkel, 2015; Spinuzzi, 2012). CSs are often referred to as *serendipity accelerators* (Moriset, 2013), born to accommodate knowledge workers, who carry out their activities by renting a workstation for a variable period, thus taking advantage of the services offered (e.g. secretarial services, Wi-Fi connection, meeting rooms, kitchen, leisure spaces, training and coaching courses, babysitting) (Spinuzzi, 2012).

The first CS in Italy opened in 2008. In 2020, Italian Coworking recorded the presence of 700 spaces in the country.[1] Of these, 119 were located in Milan since coworking is a predominantly urban phenomenon. However, temporary closures during the first lockdown and those that followed at the end of 2020 have sometimes resulted in permanent shutdowns. Lo Russo and Mariotti (2022), in their recent study on coworking spaces in Italy, found that more than 100 spaces closed for the

[1] https://www.italiancoworking.it/i-numeri-del-coworking-in-italia/

I. Mariotti (✉) · F. Ciccarelli
DAStU, Politecnico di Milano, Milan, Italy
e-mail: ilaria.mariotti@polimi.it; francescachiara.ciccarelli@polimi.it

I. Pais
Università Cattolica del Sacro Cuore, Facoltà di Economia, Milan, Italy
e-mail: ivana.pais@unicatt.it

J. Merkel et al. (eds.), *Coworking Spaces*, https://doi.org/10.1007/978-3-031-42268-3_11

entire year 2020, with many waiting to reopen in 2021. Furthermore, 66 spaces already closed permanently in 2020, and several others followed in 2021. Nevertheless, the closure of spaces has been partially offset by investments planned and/or initiated before the pandemic, leading to the birth of 36 new coworking spaces in 2020. The pandemic has also accelerated the attractiveness of suburban, peripheral and rural areas: by the end of 2020, around 20 new spaces have opened in Milan's urban belt, while the Southern regions recorded 13 new spaces located in cities under 100,000 inhabitants (Lo Russo & Mariotti, 2022). These areas have become more attractive for new workplaces that can host remote workers and digital nomads. In 2021, 46% of the coworking spaces surveyed by Lo Russo and Mariotti (2022) hosted digital nomads and 65% remote workers. Although there are few studies on the migration of knowledge workers from northern Italy or Europe toward the South of Italy or the Italian Inner Areas, available research suggests that these remote workers are more likely to be male, single or part of a couple without children (Di Matteo et al., 2021).

During lockdowns, many workers in the public and private sectors were encouraged or forced to work remotely from their own homes, while others opted for new spaces, such as CSs, sometimes drawing on the support of corporate or public policies. Indeed, Covid-19 has altered the profile of CS users, with an increase in the number of dependent employees working remotely. In addition, there has been a rise in the number of CSs in areas of the country marked by depopulation and economic decline, especially in inner areas that are characterised by their significant distance from urban centres offering essential services in terms of healthcare and education (Barca et al., 2014) – and especially in the South of Italy. In the last decade, more than 500,000 people, primarily young people of working age, migrated from the South and the islands to the North-Central parts of Italy, especially to Lombardy and Emilia Romagna (ISTAT, 2021). According to the official national statistics, this phenomenon can be considered skilled migration since over 40% of people over 25 who emigrate hold a university degree. However, at the beginning of the lockdown periods, in the early months of 2020, the situation slightly changed. The net gain of residents in Milan decreased drastically from 3% to 0.3%, while the Southern Italian provinces saw a reduction in the number of residents who left during this year (ISTAT, 2022).

Although it is not possible to talk about an actual urban exodus to the countryside, it was estimated that around 45,000 dependent workers of large enterprises could be classified as "South workers" – individuals working remotely from the South of Italy for a company headquartered in the North or abroad (SVIMEZ, 2020). Indeed, many workers, who had previously moved to other parts of Italy or abroad, returned to their hometowns and cities in the South. In some cases, the decision to leave cities and move to less densely populated areas can be seen as a form of privileged mobilities, driven by a desire for a lifestyle change and the search for more idyllic surroundings (Benson & O'Reilly, 2009; Cohen et al., 2015), which has been the case for several second-home owners in Italy. In other cases, these counter-urban movements may be framed as *crisis-led* (Gkartzios, 2013), as individuals seek to escape the diseconomies of big cities, such as high rental costs, the small dimension of dwellings and reduced access to green areas.

The Italian association "South Working – Lavorare dal Sud" (SW-LdS, www.southworking.org) has played a key role in promoting the idea that people working remotely might be able to live, even if for just limited periods, in the South of Italy and inner areas, working from the so-called Presidi di Comunità – *community garrisons* (Mirabile & Militello, 2022). These spaces are defined as "not just simple workplaces other than the traditional office, but spaces for collaboration and innovation in which to strengthen ties between new and existing communities" (Mirabile & Militello, 2022, p. 93).

As a result of the changes brought by the Covid-19 pandemic, local administrations have significantly increased their attention to these spaces (Manzini Ceinar & Mariotti, 2021; Mariotti et al., 2022a, b; Vogl & Akhavan, 2022). This trend is particularly pronounced in areas outside urban centres where CSs have been more widely used as policy tools to compensate for the lack of private initiative, which is due to the relatively low demand for such services in these areas. Indeed, CSs in rural areas and other parts of Europe are frequently subsidised and hosted in public converted buildings (Tomaz et al., 2022).

Since the emergence of CSs, many public administrations in Italy, especially at the local level, have recognised their public value, promoting them through policies that fall into three areas: (i) labour-oriented, (ii) social innovation and (iii) local economic development policies (Mariotti & Pais, 2022). The last category includes the promotion of public coworking spaces. The Italian case is particularly interesting since the first cases of public intervention date back to 15 years ago, when the coworking phenomenon was still in its infancy. Yet, over time these experiences were gradually abandoned and were rediscovered, in different forms, only during the Covid-19 emergency.

This chapter aims to analyse the case of these new types of public working spaces and the Presidi di Comunità, which are initiated by private citizens who get some form of public support for their establishment. The collective aim behind these initiatives is to stimulate the local creative ecosystem and establish a fruitful relationship between remote workers and local communities. The chapter is composed of four sections. The introduction is followed by a description of the three typologies of policies supporting CSs usually adopted in the Italian context, as well as the changes observed since the start of the Covid-19 pandemic. Section 3 presents several cases of Presidi di Comunità opened before the pandemic and established through the collaboration between private citizens and public institutions, based on interviews with founders and managers. This section describes the rationale behind their opening, their business strategy, the services provided, the type of users and their role in the local context. Conclusions and suggestions for further research follow.

The Role of Policies Promoting Coworking Spaces in Italy

Collaborative workspaces, including CSs, have become part of policies at different levels (ranging from the European to the municipality level; see Avdikos & Merkel, 2020) with the goal of promoting local development, employment and social innovation. However, the available literature on which types of spaces are funded and how they are funded remains relatively scarce. In a recent analysis of the policies supporting CSs worldwide, Avdikos and Papageorgiou (2021) found that most cases were funded to enable entrepreneurial growth and contribute to local and regional development, with policy objectives aligning with the second wave of neoliberal model of coworking (Gandini & Cossu, 2021). On the other hand, policies directed at reducing commuting and promoting these spaces as a tool for innovating the public sector or promoting social infrastructure were less common (Avdikos & Papageorgiou, 2021). Before the pandemic and the emergence of CSs in Italy, these spaces have been partially supported by public administrations through actions related to three policy areas: (i) labour-oriented, (ii) social innovation and (iii) local economic development policies (Mariotti & Pais, 2022).

Labour-oriented policies supporting CSs have taken the form of individual vouchers issued directly to workers, mainly freelancers. The first initiatives were promoted in 2013 by the City of Milan and the Tuscany Region to support a particularly vulnerable category of workers while promoting the mapping of CSs and raising awareness of their existence. Thus, these actions indirectly supported the diffusion of these spaces (Tajani, 2021). During the Covid-19 crisis, freelance workers were severely affected by the freezing of consultancy activities and received less public income support than employees and entrepreneurs (Gerosa et al., 2021). Nevertheless, the voucher model was not the primary policy to support these workers, with the exception of the Apulia region. In 2020, the regional administration allocated 200,000 euros in vouchers addressed to freelancers and self-employed to use CSs, fablabs and makerspaces.[2] Recognising the heightened risk of social isolation due to the Covid-19 pandemic, the administration provided vouchers of the value of maximum 3000 euros to independent workers to be spent within the network of collaborative workspaces located in the region. Such actions demonstrate an acknowledgement of the value of these spaces in providing affordable workplaces for precarious freelancers (Merkel, 2019; Pacchi & Mariotti, 2021). Following Avdikos and Papageorgiou's (2021) reasoning, these initiatives seem to align with the first wave of coworking (Gandini & Cossu, 2021), given their focus on supporting independent workers through physical and social infrastructure.

The second strand of public intervention is oriented toward *social innovation* through partnerships with the private social sector. In this case, public-private collaboration goes beyond financial support and often involves third-sector organisations managing public spaces. CSs become an integrated service within hybrid

---

[2] https://www.regione.puglia.it/web/competitivita-e-innovazione/-/voucher-co-working-makerspace-fablab

spaces that support initiatives oriented to specific objectives but of collective interest, such as family-work reconciliation, ecological transition and social entrepreneurship. During the pandemic emergency, the co-design logic underlying these experiences facilitated the reorganisation of the services offered to respond to new needs (Mariotti et al., 2021, 2022b). An example of this type of intervention is the Municipality of Vimercate, located in the Lombardy region. In 2018, the municipality signed an agreement with the cultural association Sloworking, which promotes "work at the pace of life", to redevelop a disused public space. In this case, the space acted as a physical infrastructure for the community that had already formed around the association and its projects. According to the founder we interviewed: "the people who come here are aware that they enter a place where they do not simply find a workstation but a group of active and socially engaged people". During the Covid-19 pandemic, the space registered an increase in the number of male employees compared to the traditional users, namely, female freelancers. Moreover, thanks to a project that involved the municipality, the space offered an accessible summer centre in the coworking courtyard, open to the coworkers' children and the children living in the neighbourhood. This example shows how such spaces may adopt hybrid forms by including both activities and services open to the broader local community and tailored services for specific users. This approach reinforces the economic sustainability of the project and allows the space to take on a more comprehensive social function.

The third strand of intervention is more explicitly oriented toward *local economic development* and involves the establishment of coworking facilities with the specific objective of revitalising, repopulating and reactivating peripheral areas, in both urban and rural contexts. Civico Trame, a community hub located on the outskirts of Lamezia Terme in the province of Catanzaro (Calabria region, Southern Italy), is an example of this third strand of public policies supporting CSs. The project was initiated in 2015 involving two local grassroots organisations actively fighting against the mafia. These organisations were entrusted with the regeneration of a building that was meant to function as a senior centre 10 years earlier but was never put into operation. As of today, this multifunctional space is equipped with a library; an area dedicated to coworking and studying; an event's space for training courses, workshops and cultural activities; and a shared kitchen/café area. The space primarily hosts students but is also open to potential remote workers looking for a place to work. Spaces and services are managed directly by the associations and volunteers, and access to the spaces is free, subject to an annual membership fee of 10 euros. Among the activities promoted by the space, there is CivicUP, a project enhancing active citizenship and participation in the local community, especially for young people aged 16–26 years.

Many CSs opened with the aim of fostering local development are owned and, at times, managed by public authorities, in order to achieve this policy objective. While most CSs in Italy and other countries are privately owned[3] (such as in Finland,

[3] https://www.italiancoworking.it/i-numeri-del-coworking-in-italia/

see Kojo & Nenonen, 2016; or in France and Spain, see Capdevila, 2017), public coworking spaces have a distinct social value. Although private CSs may organise open activities and events that serve an important social function, they are not equivalent to publicly accessible social infrastructures (Avdikos & Merkel, 2020; Klinenberg, 2018). As Lorne (2020) suggests, being open does not necessarily mean being public, and not necessarily CSs' benefits are equally accessible and enjoyable for everyone. On the contrary, this is one of the most significant potentials for public coworking spaces.

The first case of public CS in Italy was recorded in 2011 in Veglio, a municipality of 500 inhabitants in the province of Biella in the Piedmont region. This was made possible, thanks to funding from the Permanent Secretariat of the Alpine Convention to combat the depopulation in Alpine areas and reduce commuting. However, this experience came to an end after some time, and only a few others were launched in the following years. This was mainly due to difficulties on the part of public administrations in facilitating the development of the communities that are supposed to arise in these shared spaces.

Yet, the Covid-19 pandemic provided an opportunity for the emergence of new public workspaces, particularly in Southern Italy, linked to the South Working initiative. Rather than being part of top-down policies promoted directly by administrations as it happened a decade ago, these newly opened coworking spaces take the form of bottom-up initiatives coming from private citizens and supported by public administrations. As this is the most significant novelty in the field of coworking practices and policies in Italy as a result of Covid-19, we will engage with an in-depth discussion of this phenomenon in the next section.

The Revival of Public Coworking: The Role of Presidi di Comunità

Public coworking spaces come in a variety of forms and cater to different user needs, ranging from freely accessible spaces with single or shared desks to hybrid spaces that offer coworking in combination with other activities and services (Avdikos & Merkel, 2020). Indeed, public coworking spaces can be hosted by universities, research institutes and libraries (Wyatt et al., 2015). In several countries like Finland, Norway, Canada, the Netherlands, the UK and Australia, public libraries have been transformed into multifunctional spaces (Bilandzic & Foth, 2013, Di Marino & Lapintie, 2020; Mariotti et al., 2021), including also spaces for work. Public coworking spaces are also being opened to reduce commuting and congestion, especially when located in rural and peripheral areas of towns and cities, with the idea of improving workers' well-being and work-life balance. This approach recognises that the time saved by workers who no longer have to travel to work can be used for spending time with family and engaging with leisure and community activities (Houghton et al., 2018).

In contrast to 10 years ago, when local administrations promoted public coworking spaces, during lockdowns, the initiative has been taken by private citizens, by establishing Presidi di Comunità. This approach is not entirely novel, as entrepreneurs have, already in the past, taken the lead in promoting economic development and attractiveness in their regions. However, during lockdown periods, it was the employees of companies with remote headquarters, rather than entrepreneurs, who actively proposed these initiatives.

This section focuses on the South Working network and the Presidi di Comunità presenting some examples of their activities. These cases have been selected based on their significance, rather than to represent the entire universe of public coworking spaces. The founders and managers of these cases were interviewed during the spring and summer of 2022.

The South Working network emerged during the pandemic, providing structure and networking opportunities for isolated coworking spaces that were sprouting up in different parts of the country, especially in the South of Italy. In March 2020, the Association "South Working" was founded by a group of young professionals and students linked to the "Community Global Shapers Palermo Hub" and united by the condition of expat or out-of-towner. South Working promotes the idea that location-independent workers can live, even if just for limited periods, in the South of Italy and the Italian inner areas, working in the Presidi di Comunità that provide stations for remote workers and digital nomads (Mirabile & Militello, 2022). At the same time, these spaces may offer services for the wider local community, rather than targeting specific groups of users, in line with models of CSs that are more open and community led (Avdikos & Iliopoulou, 2019; Avdikos & Merkel 2020).

The main objectives of South Working are summarised below (Mirabile, Militello, 2022; Mariotti et al., 2022a): (i) promote coworking as a new way of working, promoting remote habits among entrepreneurs, freelancers and employees; (ii) create synergies with other peripheral areas; (iii) attract remote workers and retain talents (that moved to work in the north of Italy or abroad) to rural areas; and (iv) contribute to the economic, social and territorial cohesion, thus reducing inequalities.

South Working collaborates with over 50 municipalities in Italy. By signing a memorandum of understanding, municipalities and other local authorities can become part of the South Working network, with the aim of favouring remote workers' return in the areas they govern. In addition to working with single municipalities to establish public coworking spaces, South Working also collaborates with pre-existing private spaces. As of mid-2022, the South Working network included 230 public and private spaces. The network has over 2000 registered workers, and more than 50 companies have agreed to allow their employees to work remotely from the South and have signed a protocol or letter of intent with the Association.

In *Castelbuono*, a town of 8500 inhabitants in the province of Palermo, Sicily, a group of workers who had temporarily returned to their village during the first lockdown initiated a coworking space. The interview with one of the founders highlights that one of the main drivers for remote workers' activation was their personal need for a physical and social infrastructure:

> We do it because this is our home, somewhere else we would not, we are all volunteers, we are not interested in money, we were interested in creating such a space so that we could use it and so that we could develop dynamics between us. A start-up has already been created in these rooms.

Moreover, remote workers were highly involved in the establishment of the space and shared their professional skills for the development of the project:

> We said to the mayor: we'll do everything. We have a computer engineer, a web designer, a lawyer, a project manager… just give us the chance to do it. And he said yes, also because he knows and trusts us.

The mayor made unused spaces available in the naturalistic museum and the civic museum with the idea of contributing to local development by leveraging the human capital that returned to the village during the Covid-19 crisis:

> We are demonstrating that villages and territories can be enhanced by bringing to the territories the excellence of young people who have gone away and who can now give a contribution to our territories.

During the first lockdown, a coworking facility in *Tursi* (Basilicata) has been opened in a former seventeenth-century convent, home to permanent exhibitions. The initiative aimed to bring the historic centre back to life. Also in this case, we see a direct voluntary involvement of the founder, not only for satisfying a personal need, but to promote local development:

> The building has been renovated but it is not used every day; with this initiative, we aim to bring this building and the historic centre back to life. [...] I have no intention of making profits from this activity because there is no way to do it through a CS in a small town. It is more of an innovative service offered to the community; I invest in it. I think we should promote sustainable tourism by attracting digital nomads.

Along the same lines as the point made by the mayor of Castelbuono on the importance of attracting and retaining local human capital, the councillor for youth policies in Tursi commented:

> The pandemic was an opportunity to have these workers again in our area. It is an opportunity to have their professionalism at the service of the community.

In Vibo Valentia (Calabria), the public library saw an increase in the number of users during the Covid-19 pandemic and decided to support the request to promote a space dedicated to coworking. The initiative was led by a young IT engineer working remotely during the Covid-19 emergency, who returned to his hometown driven by a give-back logic:

> I wanted to return to the base, in my beloved land, to give something back [...] also to give other kids the possibility of being able to choose whether to want to leave and not be forced to do so.

Such a driver is also confirmed by the head of the library who stated:

> Workers who come here not only manage to work but also commit themselves to the territory.

Similar to Presidi di Comunità, *Coworking Valsesia*, a small network of shared offices, opened during the pandemic in Piedmont's Alps. The initiative was led by a private citizen who had lived abroad for some years and was now working remotely from her hometown. She saw the possibility to open multiple small coworking spaces as a possible opportunity for the development of the valley. Seven small, shared workspaces were open to accommodate remote workers and local employees who sometimes had to travel to larger cities. Five municipalities of the Sesia Valley made several unused or underutilised spaces available for free, converting them into shared offices with desks and access to a fast Internet connection. These shared desks are attracting temporary residents, particularly second-home owners, and tourists, who can extend their stay in the area by working remotely using these facilities. Indeed, the crucial element in this experience, rather than desks, seems to be the Wi-Fi access points, given the typically limited, or even absent, network coverage in mountain valleys. The provision of digital infrastructure, a crucial issue for rural areas (Salemink et al., 2017), mirrors the role played by the first telecentres opened in rural areas decades ago, which became tools for local development when access to internet connection was even scarcer (Errichiello & Pianese, 2020).

The experiences described above share the same ingredients, taking place in economically and socially depressed areas that suffer from outmigration. One of the key findings is that remote workers have been the primary initiators of public coworking spaces during the pandemic, returning to their village of origin and needing a place to work but also strongly motivated by a give-back logic to contribute to local development. Indeed, besides being moved by personal needs, they have also offered expertise and skills to establish physical coworking spaces, applied for the necessary funds and managed social media to attract potential users. Besides using their own professional skills (e.g. lawyers provided legal advice, and computer scientists worked on the website), they have also mobilised their social networks, contributing also indirectly to the initiatives activated in the spaces. In all the cases analysed, these activities were carried out on a voluntary basis. Local governments also play a crucial role in supporting these initiatives, usually by providing public premises for their development. In this sense, establishing Presidi di Comunità also represents an opportunity for urban regeneration, repurposing and revitalising abandoned buildings, but also to encourage sustainable tourism practices, by providing visitors with the chance to extend their stay in the area.

Comparing public coworking spaces settled before the pandemic and spaces developed during the pandemic highlights significant differences. Spaces set before the pandemic followed a top-down approach, with public administrations initiating and managing them. With the outbreak of the pandemic, a new demand for hybrid spaces emerged from remote workers who, when settling back in their hometowns in peripheral areas, were hit by a need for digital and office infrastructures but also by the desire to contribute to the places they were inhabiting. Local administrations subsidised these new spaces, entrusting them to the single or groups of remote workers and, in some cases, providing maintenance services (e.g. cleaning), but without offering support for community management services.

Presidi di Comunità are inhabited not only by people who have returned to Southern Italy and work for companies located elsewhere but also by digital nomads and location-independent workers who reside for limited, but still extended, periods in less central and urbanised areas (e.g. second-home owners). Finally, such spaces may be used by employees of companies that have signed an agreement with these coworking spaces. This latter aspect is perhaps the most interesting because, according to the managers of the Presidi di Comunità, it may stimulate more investment in the region, besides guaranteeing a more appropriate workspace for remote workers.

At the time of writing (summer-autumn 2022), there is no data available about the number of *South workers* that are steadily working in Presidi di Comunità. According to Mario Mirabile, Co-founder, Executive Vice-president and Project Manager of South Working, the situation is very heterogeneous: some remote workers (whether returnees or not) moved to the Southern and inner areas of Italy, some stayed, while others went back to bigger urban centres. In general, for Presidi di Comunità reaching a large number of users is the biggest challenge. Still, some of them managed to establish a stable user base, even though granular information on the characteristics and the number of these users is not available yet. In particular, some spaces managed to attract a more significant and stable number of users, thanks to the media coverage they received. Moreover, during the several phases of the pandemic, some spaces seem to have adapted to the changes by attracting different users in different strands of time. For instance, some Presidi di Comunità have started to host local students when workers returned to their offices. These spaces opened during the pandemic cannot be considered as a kind of *emergency infrastructures*, to be dismissed at the end of the pandemic. On the contrary, and in line with the available literature, spaces located in rural and peripheral areas seem to have demonstrated to be remarkably resilient (Görmar, 2021; Tomaz et al., 2022; Vogl & Akhavan, 2022).

Conclusion and Further Research

During the pandemic, several knowledge workers have shown their willingness to work outside big cities, returning to their hometowns or pursuing a better quality of life in peripheral and rural areas. The Covid-19 pandemic has triggered changes in the balance among labour, social innovation and local economic development policies supporting coworking spaces. In particular, the pandemic resulted in the emergence of public coworking spaces in Southern Italy, following successful examples in Northern Europe.

This chapter analysed the public policies supporting coworking spaces in Italy in the last 15 years, identifying three typologies of policies based on their objectives and targets. The first group of policies directly supports independent workers; the second provides support to CSs favouring social innovation; the third, with the objective of favouring local development, foresees also the opening of public coworking spaces. While in the past this last strand was characterised by a top-down approach, in the last years this trend was reversed, with single or multiple workers taking initiative in the establishment of coworking spaces. In this context, the case of Presidi di Comunità emerged as a distinct phenomenon during the pandemic period.

The analysis of the cases studied highlights the potential of these spaces to serve both remote workers and local development while simultaneously underscoring the importance of direct citizens' participation and of the integration with other local policies to support long-term (re)location choices. Presidi di Comunità can be considered as an interesting tool for the place-based policy approach, since they are built on the embedded local knowledge and aim to enhance the "sense of community" (Barca et al., 2012). The direct participation of private citizens and the enhancement of local resources for local development, including human capital, are indeed key features of place-based policy approaches (Barca et al., 2012; Bentley & Pugalis, 2014). Furthermore, the establishment of these spaces in peripheral areas may tap into the underutilised potential of these places (Barca, 2009).

Two issues remain open. Firstly, further research is needed to better understand daily life within these spaces. Understanding how wide and varied is the user base of Presidi di Comunità is surely one first necessary step. Moreover, it is necessary to explore how common experiences of collaboration and innovation are and if they are comparable to those in urban areas and private coworking spaces. In this regard, the available literature on coworking spaces has very much underlined the role of coworking hosts (Ivaldi et al., 2018; Merkel, 2015, 2019). Investigating this aspect may be crucial to understand whether and how coworking communities are forming. In particular, it is essential to delve deeper into understanding the role of the promoters of Presidi di Comunità to see whether they also act as community hosts or whether someone else is taking this role. Given their centrality, it is important to analyse their role in the transition from the nascent phase to the institutionalisation of spaces and what happens if (for a variety of reasons) they reduce their level of commitment or leave the space. Furthermore, equally significant is determining whether and in what ways those who have returned to their urban offices and homes continue to collaborate with the Presidi di Comunità. Specifically, they may contribute to managing initiatives that seek to advance the local community, such as organising training programs for youth, fostering local entrepreneurship and other similar activities.

The second issue concerns the potential effects of the establishments of public coworking spaces on repopulation and revitalisation processes in depressed peripheral and rural areas. Indeed, peripheral and rural areas suffer from diseconomies related, for instance, to poor access to services, including transport, health and education. This explains why a *reverse brain drain* (Bakalova et al., 2021) did not occur as some expected. After the end of lockdowns, remote workers seem to be (at least partly) returning to their corporate offices and are using CSs in inner areas mostly for temporary stays. This has not led to the closure of Presidi di Comunità, which, on the contrary, are still essential infrastructures for attracting new temporary residents.

Presidi di Comunità represent *local collective competition goods* (Crouch et al., 2004), following the literature on local development, which introduced this concept to indicate the goods and services made available in a specific territorial context. On the one hand, the presence of a coworking space represents a resource available to the people, companies and organisations who inhabit it (Pais, 2022). However, on the other, coworking spaces need other local collective goods to build an attractive ecosystem for workers, especially to foster repopulation processes. On the first front, that of policies supporting coworking as local collective goods, the pandemic has stimulated the activation of public administrations that, with limited investment, have been able to support private initiatives. When considering more structural measures such as the investment in local services in inner areas needed to improve residents' quality of life, a lot still needs to be done. As the cases studied show, without structural investments in depressed areas, coworking spaces such as Presidi di Comunità may favour inward mobilities that are only temporary. Durable repopulation and regeneration processes necessitate a more consistent effort and political will.

The National Recovery and Resilience Plan (PNRR) has the potential to play a role in this context, but its implementation may face limitations, particularly in small municipalities of peripheral regions due to a lack of human resources. The Plan foresees the provision of resources to promote the expansion of broadband, the recruitment and development of human capital, with particular attention to women and remote working in public administrations. Expanding broadband is an unavoidable precondition for enhancing measures to support remote working, participation in the digital economy, the promotion of entrepreneurship and the training of young people. Finally, the PNRR also envisages the creation of Territorial Poles for recruitment, training, coworking and remote working. These initiatives could build on the bottom-up and place-based approach characterising Presidi di Comunità and other public coworking spaces activated during the Covid-19 pandemic, to support digitalisation and the development and retention of human capital in peripheral areas and eventually contribute to a more balanced regional development.

Acknowledgements The research was conducted in the framework of the COST Action CA18214 Project "The geography of New Working Spaces and the impact on the periphery", funded by the Horizon 2020 Framework programme of the European Union (project website: http://www.new-working-spaces.eu/; European Union Website: project website: https://www.cost.eu/actions/CA18214) and of the CORAL ITN-Marie Curie (https://coral-itn.eu/about-2/). We want to thank Maria Chiara Negri Da Oleggio and Andrea Chiruzzi for their support during the interviews.

References

Akhavan, M. (2021). Third places for work: A multidisciplinary review of the literature on coworking spaces and maker spaces. In I. Mariotti, S. Di Vita, & M. Akhavan (Eds.), *New workplaces—Location patterns, urban effects and development trajectories: A worldwide investigation* (pp. 13–32). Springer. https://doi.org/10.1007/978-3-030-63443-8_2

Avdikos, V., & Iliopoulou, E. (2019). Community-led coworking Spaces: From co-location to collaboration and collectivization. In R. Gill, A. C. Pratt, & T. E. Virani (Eds.), *Creative hubs in question* (pp. 111–129). Springer. https://doi.org/10.1007/978-3-030-10653-9_6

Avdikos, V., & Merkel, J. (2020). Supporting open, shared and collaborative workspaces and hubs: Recent transformations and policy implications. *Urban Research & Practice, 13*(3), 348–357. https://doi.org/10.1080/17535069.2019.1674501

Avdikos, V., & Papageorgiou, A. (2021). Public support for collaborative workspaces: Dispersed help to a place-based phenomenon? *Local Economy, 36*(7–8), 669–682. https://doi.org/10.1177/02690942221074941

Bakalova, I., Berlinschi, R., Fidrmuc, J., & Dzyuba, Y. (2021). *COVID-19, working from home and the potential reverse brain drain*. GLO Discussion paper series 845. Global Labor Organization (GLO). https://www.econstor.eu/bitstream/10419/234120/1/GLO-DP-0845.pdf

Barca, F. (2009). *An agenda for A reformed cohesion policy: A place-based approach to meeting European Union challenges and expectations*. Independent report, Prepared at the request of the European Commissioner for Regional Policy, Danuta Hübner. European Commission.

Barca, F., McCann, P., & Rodríguez-Pose, A. (2012). The case for regional development intervention: Place-based versus place-neutral approaches. *Journal of Regional Science, 52*(1), 134–152. https://doi.org/10.1111/j.1467-9787.2011.00756.x

Barca, F., Casavola, P. & Lucatelli, S. (2014). *Strategia nazionale per le Aree interne: definizione, obiettivi, strumenti e governance*. Materiali Uval n. 31.

Benson, M., & O'Reilly, K. (2009). Migration and the search for a better way of life: A critical exploration of lifestyle migration. *The Sociological Review, 57*(4), 608–625. https://doi.org/10.1111/j.1467-954X.2009.01864.x

Bentley, G., & Pugalis, L. (2014). Shifting paradigms: People-centred models, Active regional development, space-blind policies and place-based approaches. *Local Economy, 29*(4–5), 283–294. https://doi.org/10.1177/0269094214541355

Bilandzic, M., & Foth, M. (2013). Libraries as coworking spaces: Understanding user motivations and perceived barriers to social learning. *Library Hi Tech, 31*(2), 254–273. https://doi.org/10.1108/07378831311329040

Capdevila, I. (2017). A typology of localised spaces of collaborative innovation. In M. van Ham, D. Reuschke, R. Kleinhans, C. Mason, & S. Syrett (Eds.), *Entrepreneurial neighbourhoods: Towards an understanding of the economies of neighbourhoods and communities* (pp. 80–97). Edward Elgar Publishing. https://doi.org/10.4337/9781785367243.00013

Cohen, S. A., Duncan, T., & Thulemark, M. (2015). Lifestyle mobilities: The crossroads of travel, leisure and migration. *Mobilities, 10*(1), 155–172. https://doi.org/10.1080/17450101.2013.826481

Crouch, C., Le Galès, P., Trigilia, C., & Voelzkow, H. (2004). *Changing governance of local economies: Responses of European local production systems*. Oxford University Press.

Di Marino, M., & Lapintie, K. (2020). Exploring multi-local working: Challenges and opportunities for contemporary cities. *International Planning Studies, 25*(2), 129–149. https://doi.org/10.1080/13563475.2018.1528865

Di Matteo, D., La Regina, R., Mariotti, I., & Militello, E. (2021). Quali determinanti per il South Working? Una nuova proposta di sviluppo per il Sud, le aree Interne e il Paese. *Rivista Economica del Mezzogiorno, 4*, 678–701. https://doi.org/10.1432/104472

Errichiello, L., & Pianese, T. (2020). Toward a theory on workplaces for smart workers. *Facilities, 38*(3–4), 298–315. https://doi.org/10.1108/F-11-2018-0137

Gandini, A., & Cossu, A. (2021). The third wave of coworking: 'Neo-corporate' model versus 'resilient' practice. *European Journal of Cultural Studies, 24*(2), 430–447. https://doi.org/10.1177/1367549419886060

Gerosa, A., Manzo, C., & Pais, I., (2021). Il lavoro condiviso: La trasformazione degli spazi di coworking durante l'emergenza Covid-19. In Ambrosianeum, *MILANO 2021:Rapporto sulla città. Ripartire: il tempo della cura* (pp. 100–114). Franco Angeli.

Gkartzios, M. (2013). 'Leaving Athens': Narratives of counterurbanisation in times of crisis. *Journal of Rural Studies, 32*, 158–167. https://doi.org/10.1016/j.jrurstud.2013.06.003

Görmar, F. (2021). *Collaborative workspaces in small towns and rural areas. The COVID-19 crisis as driver of new work models and an opportunity for sustainable regional development?* (Working paper).

Houghton, K. R., Foth, M., & Hearn, G. (2018). Working from the other office: Trialling co-working spaces for public servants. *Australian Journal of Public Administration, 77*(4), 757–778. https://doi.org/10.1111/1467-8500.12317

ISTAT. (2021). *Iscrizioni e Cancellazioni Anagrafiche della Popolazione Residente Anno 2019.* https://www.istat.it/it/files/2021/01/REPORT_MIGRAZIONI_2019.pdf

ISTAT. (2022). *Iscrizioni e Cancellazioni Anagrafiche Anno 2020.* https://www.istat.it/it/files/2022/02/REPORT_MIGRAZIONI_2020.pdf

Ivaldi, S., Pais, I., & Scaratti, G. (2018). Coworking(s) in the plural: Coworking spaces and new ways of managing. In S. Taylor & S. Luckman (Eds.), *The new normal of working lives: Critical studies in contemporary work and employment* (pp. 219–241). Springer. https://doi.org/10.1007/978-3-319-66038-7_11

Klinenberg, E. (2018). *Palaces for the people: How social infrastructure can help fight inequality, polarization, and the decline of civic life*. Crown Publishing.

Kojo, I., & Nenonen, S. (2016). Typologies for coworking spaces in Finland – What and how? *Facilities, 34*(5/6), 302–313. https://doi.org/10.1108/09564230910978511

Lo Russo, M., & Mariotti, I. (2022). L'impatto della pandemia Covid-19 sulla sostenibilità e resilienza degli spazi di coworking. In F. Montanari (Ed.), *Spazi collaborativi in azione. Creatività, innovazione e impatto sociale* (pp. 71–78). Franco Angeli.

Lorne, C. (2020). The limits to openness: Co-working, design and social innovation in the neoliberal city. *Environment and Planning A: Economy and Space, 52*(4), 747–765. https://doi.org/10.1177/0308518X19876941

Manzini Ceinar, I., & Mariotti, I. (2021). Teleworking in post-pandemic times: may local coworking spaces be the future trend? *Romanian Journal of Regional Science, 15*(1), 52–76.

Mariotti, I., & Pais, I. (2022). Lavoro a distanza e nuovi luoghi del lavoro: il ruolo dei coworking pubblici in Italia. In M. Mirabile & E. Militello (Eds.), *South working. Per un futuro sostenibile del lavoro agile in Italia* (pp. 81–88). Donzelli Editore.

Mariotti, I., Di Marino, M., & Akhavan, M. (2021). The emergence of coworking models in the face of pandemic. In J. R. Bryson, A. Lauren, L. Reardon, & A. Ersoy (Eds.), *Living with pandemics: People, place and policy* (pp. 129–139). Edward Elgar Publishing. https://doi.org/10.4337/9781800373594.00021

Mariotti, I., Di Marino, M., Akhavan, M., & Capdevila, I. (2022a). The effects of COVID-19 on coworking spaces. In K. F. Zimmermann (Ed.), *Handbook of Labor, human resources and population economics* (pp. 1–20). Springer. https://doi.org/10.1007/978-3-319-57365-6_322-1

Mariotti, I., Di Marino, M., & Bednář, P. (Eds.). (2022b). *The COVID-19 pandemic and future of working spaces*. Routledge. https://doi.org/10.4324/9781003181163

Merkel, J. (2015). Coworking in the city. *Ephemera, 15*(2), 121–139.

Merkel, J. (2019). 'Freelance isn't free.' Co-working as a critical urban practice to cope with informality in creative labour markets. *Urban Studies, 56*(3), 526–547. https://doi.org/10.1177/0042098018782374

Mirabile, M., & Militello, E. (Eds.). (2022). *South Working. Per un futuro sostenibile del lavoro agile in Italia*. Donzelli Editore.

Moriset, B. (2013). *Building new places of the creative economy: The rise of coworking spaces.* Preprint retrieved from https://shs.hal.science/halshs-00914075/document

Pacchi, C., & Mariotti, I. (2021). Shared spaces or shelters for precarious workers? Coworking spaces in Italy. *Professions and Professionalism, 11*(1). https://doi.org/10.7577/pp.3911

Pais, I. (2022). Proximate future. Cities of proximity and digital platforms. In E. Manzini (Ed.), *Livable proximity. Ideas for the city that cares* (pp. 135–158). Egea.

Salemink, K., Strijker, D., & Bosworth, G. (2017). The community reclaims control? Learning experiences from rural broadband initiatives in the Netherlands. *Sociologia Ruralis, 57*(S1), 555–575. https://doi.org/10.1111/soru.12150

Spinuzzi, C. (2012). Working alone, together: Coworking as emergent collaborative activity. *Journal of Business and Technical Communication, 26*(4), 399–441. https://doi.org/10.1177/1050651912444070

SVIMEZ. (2020). *Rapporto SVIMEZ 2020: L'Economia e la Società del Mezzogiorno.* Il Mulino. Chapter 10, p. 211.

Tajani, C. (2021). *Città prossime. Dal quartiere al mondo: Milano e le metropoli globali.* Guerini e Associati.

Tomaz, E., Moriset, B., & Teller, J. (2022). Rural coworking spaces in the COVID-19 era: A window of opportunity? In I. Mariotti, M. Di Marino, & P. Bednář (Eds.), *The COVID-19 pandemic and the future of working spaces* (pp. 122–135). Routledge.

Vogl, T., & Akhavan, M. (2022). A systematic literature review of the effects of coworking spaces on the socio-cultural and economic conditions in peripheral and rural areas. *Journal of Property Investment & Finance, 40*(5), 465–478. https://doi.org/10.1108/JPIF-12-2021-0108

Wyatt, D., McQuire, S., & Butt, D. (2015). *Public libraries in a digital culture.* Report. University of Melbourne. https://apo.org.au/node/60635

Exploring Social Innovative Collaborative Workspace Assemblages: A Comparative Study from Athens

Colm Stockdale and Lorenzo Marmo

'Resilient' Collaborative Workspaces and Social Innovation

The majority of collaborative workspaces (CWS) in urban areas can be associated with a 'neo-corporate model' of flexible work (Gandini & Cossu, 2021; Johns & Gratton, 2013). These CWS mainly offer space to support coworkers in achieving their individual objectives. Rhetoric of community and collaboration are thus embedded within such CWS, but rather they demonstrate a mode of work defined as 'collaborative individualism' (Bandinelli & Gandini, 2019). Conversely, there are alternative CWS which Gandini and Cossu (2021) define as 'resilient'. These CWS in contrast aim at creating a more communitarian environment through shared goals and engagement with their local context. Brown (2017) notes how small-scale, independent CWS often interact more with their neighbourhood than larger corporates, as many are set up by locals – offering their infrastructure to be used as semi-public spaces, open to both local residents and professionals alike. In this regard, Vidaillet and Bousalham (2020) and Smith (2020) demonstrate how such CWS offer the possibility of profound societal change in a post-capitalist perspective by offering emancipatory potential for those involved in a CWS, by enabling them to engage in a diverse array of economic activities. Other CWS can be operated to facilitate grassroots innovation used by local citizens to address local social needs (Smith & Dias, 2018). With this social orientation, CWS have potential to contribute to social value creation through addressing societal needs related to sustainability, social inclusion and civic participation (Schmidt, 2019).

Spaces can act as catalysts for shifting power relations and social exclusion dynamics which can lead to the empowerment of marginalised groups, the enhancement of social capital and the recognition of cultural diversity and local innovation

C. Stockdale (✉) · L. Marmo
Department of Economic and Regional Development, Panteion University of Social and Political Sciences, European Creative Hubs Network, Athens, Greece

J. Merkel et al. (eds.), *Coworking Spaces*, https://doi.org/10.1007/978-3-031-42268-3_12

practices. Moreover, they can function as spaces of dialogue and debate among public, private and civic actors, stimulating the creation of socially innovative projects contributing to the social sustainability of neighbourhoods and local contexts (Garcia et al., 2015; Avdikos & Merkel, 2019).

In this chapter, our view on CWS lies with the 'resilient' CWS that engage in social innovation (SI) rather than the neo-corporate. SI rhetoric is often an element of the neo-corporates, with a narrative of an individual social entrepreneur being the 'changemaker' – essentially addressing social issues through market-based solutions (Bandinelli & Arvidsson, 2013). Resilient spaces in contrast can differ through a greater connection to the social economy and more community-based approaches to SI with a focus on improving social relations and collective empowerment. Such SI is demonstrated by Smith & Dias (2018), who present a makerspace in an impoverished neighbourhood of Brazil, which offers space for social inclusion where community groups can use the space for DIY solutions to the local infrastructure – from urban gardening to rainwater collectors.

This chapter aims to contribute to the literature by offering an alternative narrative to the neo-corporate CWS, through researching two spaces that are not solely spaces of entrepreneurial individualism, but rather spaces that engage with SI by addressing social needs, improving social relations and supporting socio-political transformation in their local contexts.

Social Innovation

It is important to differentiate the interpretation of SI for our study to provide clarity to what is a 'contested concept' (Marques et al., 2018). Our understanding of SI is drawn from Moulaert and MacCallum (2019), whereby we may define SI as *innovation in social relations* (Moulaert, 2013). Furthermore, we will draw on the three dimensions of SI (addressing unmet social needs, changing social relations and socio-political transformation) which will provide us with an analytical framework for our case studies (Moulaert & MacCallum, 2019). While the 'social needs' dimension is relatively straightforward, the latter two require some explanation. *Changing social relations* refers to how relationships are collaboratively built between different actors from various strands of society, altering social relations between different groups in a bottom-up fashion, while *socio-political transformation* refers to a participatory approach related to building solidarity among the dispossessed, whereby SI 'collectively empowers people (especially the marginalised) to act, not only within the existent systems and modes of governance but also towards transforming them' (Moulaert & MacCallum, 2019: 4). These dimensions are interrelated and are co-dependent on each other, i.e. through collective empowerment social relations are addressed, from addressing social needs it adds to the empowerment dimension, etc.

Assemblage Thinking

We have drawn from assemblage thinking (Deleuze & Guattari, 1987; DeLanda 2006, 2016) to explore how we may unpack how these spaces engage in SI. This section aims to outline a few principles we will use to analyse our case studies. For DeLanda (2016: 10) assemblages are social wholes:

> Interpersonal networks or institutional organisations, that cannot be reduced to the persons that compose them, and that, at the same time, do not totalize those persons, fusing them into a seamless whole in which their individuality is losts.

In this regard, DeLanda refers to the terms 'emergent properties' and 'relations of exteriority'. Emergent properties of an assemblage do not originate from the properties of its parts but from their interactions, while relations of exteriority refer to how assemblages cannot be reduced to their constituent parts – parts have autonomy and can be detached from one assemblage and attached to another. Moreover, assemblages are shaped through their component parts, which can be either or simultaneously *expressive* or *material*. Regarding CWS assemblages, Avdikos and Pettas (2021) identify material components being the infrastructure, goods, spaces and tools while expressive as skills, experiences, knowledge and finance. The combination of these (the coworking 'praxis') is fluid, resulting in a variety of exercised capacities, co-dependent on how such CWS attach and detach from other assemblages.

Furthermore, assemblages are reinforced by *territorialisation* and *coding*. Territorialisation refers to how the identity of an assemblage is stabilised both spatially (through geographic borders) and non-spatially (through how its component parts are drawn from the same homogenous set (beliefs, ideals, community, etc.)). A highly territorialised assemblage will have strong homogeneity, firm borders and intensive internal ties, while a highly deterritorialised assemblage will be more heterogeneous, with porous boundaries with loose internal ties (Delanda, 2016).

Coding reinforces territorialisation, through 'expressive' components in an assemblage that fix the identity of a whole (DeLanda, 2016). It helps to govern such social wholes through the institutionalisation or formalisation of rules, norms and conventions, while decoding refers to how such codes are broken down, and recoding subsequently sees them enter new relations. Through the concept of coding, we may consider the notion of 'social innovation' as an attempt to 'de' and 're' code certain aspects of social life (norms, behaviours, conventions, values, etc.) towards more desirable outcomes. Through combinations of the coworking praxis, it can result in significant *re/de-coding* of certain established practices – for example, how a 3D printer is mobilised, or how space is interpreted.

Both (re/de)territorialisation and (re/de)coding may act as a means of (de)stabilisation for an assemblage. As they enter new relations and new components are added, it may significantly alter its properties (DeLanda, 2016). Therefore, assemblages are constantly being (dis)assembled, always in motion, always changing and seeking new alliances, which may either or simultaneously stabilise or destabilise it (Nail, 2017). For example, a SI initiative may enter new relations with a funding body that may allow it to have a greater impact, but it may cause it to lose some of its original identity, become more commercialised and lose supporters as a result.

Furthermore, McFarlane (2009) uses AT to unpack the simultaneous global and local activities of social movements, as 'translocal assemblages'. Translocal assemblages may operate at a 'local' level to address local issues but also at the 'global' in terms of networking and influencing policy, along with issues they address. Therefore, translocal assemblages are neither local or/nor global - but simultaneously both as they transcend spatial scale. Finally, McFarlane (2009) also highlights how power in assemblages is of multiple 'co-existences', rather than from a single central governing or hegemonic force, nor is it distributed equally.

Finally, we would like to use AT to introduce a 'more-than-human' perspective towards understanding SI. In considering the agency of non-humans, we may understand SI processes not as how people come together to address social challenges, but rather how a combination of people, objects and ideas interact to address various social challenges, and what emerges from this interaction is SI – as both a material and immaterial outcome (Chen et al., 2022).

Through these principles, we aim to explore how both spaces mobilise the aforementioned components and capacities to de- and re-code the practices of CWS as they re-interpret them into their own sets of relations and in doing so alter power relations in their local contexts. In doing so, we aim to add an additional, more analytical means of examining SI phenomena.

Local Context

Our study presents two cases that are based in Athens, a geographical context deeply shaped by the effects of the 2008 economic crisis. Subsequent austerity measures are still present – primarily through reduced wages and job losses, while the delivery of social services was impacted through cutbacks in public spending (Malamidis, 2020). Such measures naturally had a lasting effect on Greek society, with more than a quarter of the population at risk of poverty or social exclusion in 2020 (Eurostat, 2022).

Despite the severe socio-economic impacts, they provided fertile ground for many solidarity initiatives to prosper in Greece, and particularly in Athens. The city has a history of solidarity initiatives such as squats and social centres in response to a weak welfare state and the increasing commodification of public space (Hadjimichalis, 2014). Austerity acted as a catalyst for proliferation of these – such as social kitchens, work collectives, social clinics and many other forms (Malamidis, 2020). During the 'refugee crisis' of 2015, social centres opened in response to the inability of the state to cater for the needs of the large flow of refugees with some offering a plurality services and spaces such as language classes, a library, cafe and pharmacy (Agustín & Jørgensen, 2018). Moreover, social clinics emerged as autonomous, self-organised collectives of people, mainly operating on a voluntary basis, who provide free primary medical pharmaceutical care to those excluded from the health system without discrimination (Malamidis, 2020). Austerity measures severely reduced funding for the health system, resulting in drug and personnel shortages (Economou et al., 2016), while patients were charged to enter hospitals (Malamidis, 2020; Teloni & Adam, 2018). The clinics were roughly provided 50% from those involved in anti-austerity social movements and 50% from more institutional actors – municipalities, the church, third sector organisations, medical associations or union of hospital doctors (Teloni & Adam, 2018).

Such cases of solidarity remain 'alternative' solutions to the politics of the city, yet nevertheless are crucial to question the status quo, while demonstrating alternative possibilities. It is this experimentation we wish to pursue in demonstrating the alternative topologies of CWS, highlighting them not only as spaces of economic and entrepreneurial activity, but rather diverse spaces of activity – both economic and social.

Methodology

This study carried out four semi-structured interviews with the members of the management team of both Romantso and Communitism, as well as with an organiser of LGBTQ+ events in both spaces. The interviews were carried out between March and April 2022. Additional information was drawn from the websites of both spaces. Furthermore, the research applied an ethnographic approach, through the participation in several events in Romantso and Communitism, as well as through dialogue with actors revolving around such spaces. Combining these two methods is aimed to provide a richer depth of understanding of the practices, processes and relationships involved and that unfold in and around both spaces. It is acknowledged, however, interviews were solely with those involved in the spaces, speaking about their relationship with their surrounding context rather than a mixture of interviewees from the spaces and their surrounding contexts.

Findings and Discussion

The Case Studies: Romantso and Communitism

Both spaces are located in the centre of Athens in the neighbouring areas of Gerani and Keramikos, beside the central Omonia Square. Both are areas of neglect and inequality and can be characterised by their immigrant population and high concentration of drug addicts and homeless.

Romantso is a creative hub functioning as an incubator for companies within the creative industry while also organising and hosting a variety of events and activities open to the wider public. Communitism defines itself as 'an open community of creative professionals'. It revives abandoned or under-utilised cultural heritage buildings transitioning them into cultural commons entrusted and operated by active communities. Each space within Communitism becomes a community that hosts 'projects, people and ideas' (Communitism.space, 2022).

The choice of the location for Romantso (Gerani) was partly dictated by the desire to positively change the neighbourhood around Omonia Square, which was considered 'the most dangerous area in Greece and the first official ghetto' (Manager of Romantso, Interview). It was chosen for two main reasons. Firstly, to tackle the abandonment of buildings in the city and secondly for the historical importance it has for the creative people of Athens and for the urban culture, given the former presence in the building of the famous Greek magazine *Romantso* that operated until 1990.

From the outside, the building of Romantso maintains the original signage from the former homonymous magazine, while inside it is modern and sleek with an industrial design. It has a quality sound system, office space, a bar and space for conferences and events, all to give it a professional, although still underground aesthetic.

Communitism, on the other hand, takes a more DIY approach, where most of the decorating, building and repairing is done by themselves rather than employing professionals. It resides in a crumbling neo-classical building, with tables constructed from pieces of scrap wood and toilets that double as an art piece. Although the material elements may portray a less professional environment, the space nevertheless has many expressive elements which add its aesthetic and attract artists and social economy actors through this alternative, underground and 'non-consumerist' style.

At a first glance, both spaces contrast quite a lot in terms of their material and expressive components. However, when one takes into account their relations, we can see they connect to many similar networks/assemblages. Both reside in the centre of Athens and engage in the creative and cultural industries. Both CWS take actions in their neighbourhoods aimed at improving and addressing inequalities, which may be linked to the state through, for instance, immigration policies and the lack of welfare measures.

Furthermore, they are endowed with what Avdikos and Pettas (2021) describe as having 'porous' boundaries to their CWS assemblage, leaving them 'open and inclusionary for new members, new materials, new ideas, social and open innovation, etc' (p. 46). Regarding Communitism, the use of the space shifts and adapts depending on the needs of its users. It is registered as an association, so operates through the presence of different autonomous groups, mainly artists, sharing the same building. While autonomous, each group has referents from the association, who manage the space. The role of the association is to support and, when available, provide funding for some projects. Around the groups, there is a network of interns, volunteers, temporary artists and students. These ephemeral relationships result in a constant shift in the dynamics of the space, which appear to be some of their core values, leading to new collaborations and interactions. Despite making long-term planning difficult, these dynamics create what one member calls 'symbiotic relationships'.

> [User] is doing drum lessons, bringing in musicians. This is what I call symbiosis, she needs a space where to start her drum lessons, and we can use some more people that are part of the music group for other events. It's a perfect match. Communitism member, Interview

Similarly, the use of space in Romantso can vary each day. However, it can be considered as a more formal CWS; its membrane oscillates between being inclusionary and exclusionary. It provides private rented office space and hosts official conferences and events. Although it hosts neighbourhood events for the migrant community, it simultaneously hosts wine tastings, art exhibitions and electronic music performances – some free and some with an entrance fee. While this exclusivity and formality are essential to bringing in revenue and funding its activities, it appears contradictory when it comes to social inclusion. While Communitism lacks this formality, and consequently revenue, its boundary is much more porous, making it more approachable for those wanting to use its space.

Social Innovation

As referenced in our literature review, from Moulaert and MacCallum (2019), we refer to three dimensions for our SI analysis: *Addressing unmet social needs; Changing Social Relations;* and *Socio-Political Transformation*. This section shall address how the spaces address social needs and change social relations. Our findings showed although the actions of the spaces did have empowerment dynamics, we would not consider them *transformative*. This will be elaborated in the discussion section.

Social Needs

As elaborated above, both CWS are located in neighbourhoods that can be considered marginalised, despite their central locations. The spaces address social needs through material improvements in the infrastructure and through the provision of free and safe spaces for various social groups, thereby aiming to remove the neighbourhoods from the margins of society.

The initial intention of both initiatives was to have a positive influence on the neighbourhood, which had both been left derelict. Romantso, when first establishing itself during the economic crisis, came across the neighbourhood:

> It was during the peak of the recession and during the immigrant situation, it was the most inflated situation of immigration here in Greece [...] There was a study by an architectural research group and the report named this area the most dangerous area in Greece and the first official ghetto. So, we took that research and started walking around the area and found

> this building. When we found it, outside it was full of garbage, you literally wouldn't be able to walk through. It was a weird situation; it was quite abnormal for the city centre. (Manager of Romantso, Interview)

Through AT we cannot understand SI as a static thing, but constantly in a process of (dis)assembling. As new component parts enter the assemblage, relations change (Nail, 2017). Through relations of exteriority, new assemblages are connected to, allowing for new capacities and possibilities. Thus, as a new actor entered into the neighbourhood assemblage of Gerani, it gained new capacities and power – as it was now able to connect with other assemblages that were previously unreachable.

> It's really interesting if you look at pictures because the place is completely different. We started by doing a lot of stuff, the building was abandoned at the time. Initially, we had to do a lot of actions, mainly to convince the landlords to rent it to us. We also did a lot of actions in the neighbourhood...Very informal process, I would pretend I was an employee of the mayor's office and approach the cleaning department to get them to remove the trash. There was also no public lighting in this part. So, we got a small public grant and got lighting, before people would not walk through this part of the city at night. (Manager of Romantso, Interview)

The ability of the founder of Romantso to communicate with the municipality and use his power to apply for grants and clean the trash on the street had a significant impact on the neighbourhood assemblage, that prior to the introduction of Romantso it did not have.

While Communitism does not engage in neighbourhood actions on the scale of Romantso, it also represents how a new component can have an effect on its neighbourhood. As stated, the mission of Communitism is to reinvigorate old neoclassical buildings in Athens. It does this twofold. Firstly, regarding the infrastructure and secondly regarding social life. Despite not having capacities comparable to Romantso in terms of power from a charismatic leader, it is endowed with different capacities, being able to leverage its relations (of exteriority) with other assemblages in the cultural industries to find other financial ways to maintain the infrastructure. The organisation uses revenues from hosting film shoots to repair the crumbling neo-classical building it inhabits. Similar film shoots also occur on their street, whereby the local shops are paid by the film studio as reimbursement for their closure.

Moreover, both spaces cater for social needs by providing a free and safe space for certain groups but do so differently. Communitism acts as an incubator of sorts for local artists, where people can experiment with low financial means and enhance their visibility. The neighbourhood has also shaped its functionality, for the past number of years it has been used by the organisers of the Metaxourgeio carnival which is known all over Athens as being one of the liveliest and busiest carnivals in the city. The organisation of carnival would not be possible without Communitism providing tools, materials and a physical space for the construction of floats.

Similarly, the neighbourhood of Gerani has affected the functioning of Romantso, allowing for a re-coding of the activities of the CWS. Initially, the two main pillars of Romantso were to facilitate the employment of creative workers and to address the abandonment of buildings in the city. However, once established in the neighbourhood, the particularities of the area influenced their activities:

> Slowly we felt that there were some functions that we should address, so we added in the original profile the social centre, which is a function that is also not included in the municipality's plans. The social centres are buildings usually owned by regional authorities across an area and they do social functions so that local communities can use those facilities for their communal needs. And no such things existed here, and then we had special requests from the different immigrants to do their gatherings, festivities, all these kinds of things. (Manager of Romantso, Interview)

With the influx of migrants into the city, a new component entered the assemblage, shifting the focus towards other inherently unaddressed needs. Initially, Romantso was perceived by the Gerani residents as a public building. This influx had impacted the social functions of the CWS, providing them with a social space to assemble and practice cultural activities.

Over time, by getting familiar with the local people, Romantso was able to further respond to the unsatisfied needs of migrants in the area, by mobilising their existing relations with social enterprises, many of whom have office space in the CWS. For example, they began a collaboration with a group, '*Generation 2.0 for Rights*', which supports second-generation immigrants from a legal perspective, helping them to obtain a Greek passport. Other services offered have included online appointments for Visa applications, facilitating video calls to their families and legal consultations during peak waves of immigration.

Communitism does not currently have a specific focus on projects involving migrants like Romantso. However, it too shifted its focus during the refugee crisis, hosting the Greek-Syrian Youth Forum. It offers space and functions as a social/cultural incubator. It is a multifunctional space, adapting according to the needs of the users. Depending on these needs, there have been music courses, workshops with refugees, rehearsals of the local historical fanfare of the neighbourhood, training of a feminist contemporary self-organised dancing team, cabarets organised by the Athenian queer scene and movie shootings to name a few:

> There are not so many places that are so much of a chameleon. [...] Communitism is open to many different things. (Member of Communitism, Interview)

Chameleon-like could be used to describe both spaces. Their flexibility and openness compared to other organisations are a key strength, allowing them to respond and adapt to the needs of the groups they often serve.

Changing Social Relations

Both CWS are changing social relations on two levels – in terms of gender and in terms of perceptions of the neighbourhood. They connect to some common assemblages, such those of the underground music scene and the LGBTQ+ community. Romantso was one of the first gender-inclusive spaces in Athens, while Communitism is known for often offering drag shows, cabarets and other inclusive events. Despite the apparent focus on LGBTQ+, they aim at offering events that transcend any gender specificity, promoting gender-inclusive values.

Both spaces were involved in the organisation of one prominent LGBTQ+-friendly event in Athens, which began in Communitism and progressed on to Romantso as it grew in size while also influencing the nightlife scene in Athens. The open nature of such events provides a bridge between the normally exclusive LGBTQ+ nightlife and 'mainstream' nightlife. By embracing inclusivity rather than exclusivity, both spaces have aided in changing the social relations in society between different groups.

Moreover, along with actions taken to address the LGBTQ+ community, both spaces have specific actions that are aimed at changing the character of the neighbourhoods and, in turn, the perceptions of local people. After settling into the neighbourhood, the management team of Romantso discovered that Athenians were referring to this area as the 'downside of the commercial area', rather than its true name – *Gerani*. They then essentially attempted to reterritorialise the neighbourhood, to try and make it more appealing by re- and de-coding certain aspects of it.

Over time, they began to communicate and re-establish the original name and aided the local restaurants, mainly managed (and only frequented) by immigrants, by translating their menus. As part of one integration project, they organised several tours of the area and businesses, bringing many Athenians (who normally would not frequent such places) to the restaurants managed by immigrants, initiating an integration process:

> Back then I was not working in Romantso, but I remember them doing a lot of migrant events, with markets and concerts. You could just go along the street and taste the food. I think they wanted to be present here, by trying to reach out to the neighbourhood. I didn't know who was doing this exactly, but I could see it was something that was active in the neighbourhood [...] I think since Romantso opened the area is more integrated and made people from the outside familiar with the neighbourhood. People who wouldn't have come otherwise. (Community manager of Romantso, Interview)

Romantso attempted to reterritorialise the formerly dirty neighbourhood full of immigrants to 'urban exotica', a multicultural place with exotic food and music. This attempt of reterritorialisation of the neighbourhood assemblage culminated with the Gerani festival, with food, music, lectures, gatherings of minorities and guided tours occurring throughout the Gerani neighbourhood. It was the outcome of participatory planning between the Gerani residents and Romantso.

Although Communitism may not attempt to reterritorialise its neighbourhood on the scale of Romantso, it does attempt to recode certain aspects of it through the immaterial and material 'reactivation' of historical buildings. Through the rehabilitation of their building in the neighbourhood of Keramikos, they aim to recode the unsavoury character of the street by restoring the building and through the many events and workshops it holds. They do not try to push out the other residents but rather recognise their coexistence – from the neighbouring brothels to the local shops and businesses managed by immigrants.

Discussion

This chapter set out to provide a new, more dynamic perspective on SI by thinking with assemblages. Rather than a single outcome, we have tried to demonstrate that SI is both a *process* (social needs, social relations) and a *practice* (socio-political transformation) (Moulaert & Mehmood, 2020). Through AT we were provided with a new lens for viewing SI, through a relational approach and through concepts such as territorialisation and coding.

Our findings, in general, focused on the ability of the CWS to address the unmet needs that their surrounding area presented them with and how they were able to change social relations in their own novel way, something which emerged through interaction with other component parts of the assemblage, enhancing its capacities. However, although the introduction of a CWS into each respective neighbourhood assemblage has altered the properties and relations of each assemblage, it has not necessarily resulted in what Moulaert and MacCallum (2019) describe as *socio-political empowerment/transformation*.

If we may understand empowerment as processes of mobilisation and change 'that improve underprivileged individuals' and social groups' ability to create and handle mental, material, social, cultural and symbolic relevant resources' (Andersen et al. 2013, p. 199), we can affirm for both CWS this empowerment occurs – through the provision of space to marginalised groups for little to no cost and allowing them to express themselves. Romantso offers its event space for free to various groups such as immigrants, LGBTQ+ groups, students and a sex workers alliance, with a similar arrangement occurring in Communitism through the use of space for no cost. In addition to offering space and equipment, Romantso supports the organisation and marketing of such events.

However, when analysing the *transformative* dimension of SI, empowerment alone is not sufficient. In fact, following Moulaert and MacCallum (2019, p. 36), it would require a step further 'in the interplay between mobilised local actors and institutions operating at wider scales of influence', through which SI would be able to empower individuals by transforming existing systems and modes of governance (Moulaert & MacCallum, 2019). Thus, empowerment needs to be institutionalised through connections with localised actors and institutions, so that their actions do not exclusively address social needs but also – through linking top-down and bottom-up action – offer alternative ideas on how these needs may be addressed (ibid.).

Returining to the social clinics, they addressed social needs through the provision of free healthcare and changed social relations by bringing medical professionals, social workers and civil society members together while working with social movements and/or formal institutions. Their transformative dimensions lay firstly in trying to shift perceptions around healthcare – that it is a social good rather than privately commodified in both medicine and service provision – and secondly through their insistence that they were not a substitute for the state but aimed to activate people to demand collectively for their rights (Malamidis, 2020; Teloni & Adam, 2018).

The failure of these CWS assemblages in enabling socio-political transformation may be due to the limits of their capacities. Both CWS may engage with SI, as they are both driven by a desire for social good. However, the extent to which they can have a transformative effect on their neighbourhood assemblage has boundaries, which can mainly be explained through their relations of exteriority.

Firstly, although both CWS are engaged in neighbourhood regeneration, it is not their sole focus. Their activities are primarily related to the creative and cultural sector. While this allows them to tackle social issues in creative and unconventional ways, it also limits the actions they can have. In fact, their main role is to provide infrastructure for creative purposes, organising festivals and workshops, rather than focusing on the provision of primary social services that address physiological needs for the most marginalised. Consequently, it is unclear what needs are really being addressed. Although this may open the CWS to critique, they are still the only social operators in the area. Therefore, their presence contributes to the improvement in the neighbourhood assemblages compared to before.

Moreover, finances are limited for both spaces, which limits the number of social actions they can have. Some of Romantso's social activities have been financed through external fundings and grants, even though a significant share of them was financially covered by the other more lucrative activities carried out by Romantso itself. When such sources of revenues reduced due to COVID-19, Romantso was forced to stop all of its social centre activities. As opposed, Communitism relies mainly on external funding to pay the rent and building restoration, such as through hosting film shoots or EU projects. The CWS do not offer a solution to the social inequalities in their respective neighbourhoods but address them in their own, creative way. Solving such inequalities is related to much broader, structural and global issues that are far beyond the capacities of the CWS.

What is clear is that an assemblage with the capabilities to provide additional capacities to realise their transformative potential to the respective assemblages is the state assemblage. The inequalities present in the areas are arguably a result of its negligence and lack of welfare services. Improved services and possibly support towards recognising the activities of the CWS could provide for further improvements in the neighbourhoods. Despite this, both spaces relate to the state differently. As discussed, Romantso has much more connections with formal networks than Communitism, making it a formal organisation with a social centre function. However, the informality of Communitism gives it the appearance of a traditional social centre, affecting the perceived legitimacy of Communitism in the eyes of the state.

These formal networks and power relations give Romantso the legitimacy to tap into the state assemblage. It is well-known to the Greek authorities – they are the only recognised official incubator for the arts in Greece and are held in high regard by the ministry of culture by being mentioned as best practice in their reports. When first arriving in the neighbourhood, they were able to acquire grants to clean up the neighbourhood. It allows them to renew their alcohol licence annually without trouble, and they were previously approached by the municipality to build a social programme, even though they were left without compensation. Such examples show the potential of the state assemblage to provide additional capacities to the neighbourhood assemblages, although they also highlight it cannot be relied upon. Communitism on the other hand has no relationship with the state and is less of a 'formal' organisation than Romantso. It is questionable whether it has an alcohol licence, and although it gives the appearance of a squat, it is officially registered as an association. While this may hinder the organisation in receiving funding, the case of Romantso shows this is not necessarily something which is guaranteed, even when officially recognised by the Ministry of Culture.

Building a relationship with the state may not necessarily be needed despite the lack of dealings with the state. Both CWS are part of separate networks of creative hubs in Europe. Here we may refer to the idea of translocal assemblages of McFarlane. Translocal assemblages are neither local nor global, but translocal. They are involved with local issues (i.e. concerns of creative hubs) while remaining global (i.e. through influencing EU policy, creating funding programmes, etc.). Both benefit from their translocal network by bypassing the need to rely on the state assemblage through accessing funding opportunities, information exchange and networking. It allows it to connect to international programmes such as Erasmus+ and international artists who can enact different projects in the space. Through their relations of exteriority, the CWS can bypass their constrained local capacities by attaching to their respective translocal assemblage.

Despite the lack of external finance from the state, they have persevered in providing various social services for those who may need them. However, although both organisations may be doing their best to reduce inequalities, it is clear there exist broader structural issues in Athens, and on the national level, that can be tackled only systematically, making it impossible for a single organisation to address them alone. As Moulaert and MacCallum acknowledge, however, SI is not just about solving day-to-day problems as a result of the dysfunctional welfare state, rather it is a means of criticising it and offering ideas at an institutional level on how things can change.

Conclusion

For the CWS to have the socio-political transformative power that SI can have, they must go a step further to enact real change. We have seen that both CWS are limited to what they can do in terms of their finances, with COVID-19 putting a stop to Romantso's social centre almost entirely. To overcome this, they must link with other actors to increase their capacities.

Returning to the social clinics, they were formed by a mixture of informal and formal institutional actors, such as social movements and municipalities. While we have above outlined the limitations of interacting with the state, collaboration with the municipality offers some potential. Municipalities acted as infrastructure providers for the clinics or aided in cost reduction by paying their rent or utilities or through EU funds which allowed the clinics to focus on their primary activities (Malamidis, 2020, Teloni & Adam, 2018).

Due to the lack of a social centre in Gerani being the reason for Romantso's actions, it could be fruitful (if possible) to partner with the municipality to set up a new social centre in the neighbourhood, through a participatory approach with the various groups that already use its space and services. It could act as a node or a broker between the other assemblages it attaches itself to, with potential to draw EU funds from its European network, creative input from its current users and the municipality for the new building and some of the running costs. This would allow it to be less reliant on the income generated by Romantso while still allowing for a creative approach. This approach may not suit Communitism, as they are more concerned with the recoding of their specific building (for now, at least) rather than territorialising their neighbourhood.

In this short chapter, we have attempted to demonstrate an alternative perspective on CWS, by examining their relationship with their local contexts rather than solely the relationships inside their walls. In tandem, we have aimed to address a gap in the CWS literature by showing how CWS engage in SI. In this regard, we base ourselves on the SI discourse in its opposition to economic innovation, interpreting it as being innovation in social relations by addressing human needs, rather than ethical economic innovations – something which is clear by the actions of each space. In addition, we have used assemblage thinking to provide us with new tools in analysing SI, through concepts such as the exteriority of relations, coding and territorialisation. Regarding the agency of non-humans, we have tried to express how the characteristics of the neighbourhoods (e.g. trash on the street and its subsequent removal) played a role in the emergence of socially innovative actions, but this must be explored further in detail.

Both spaces provide examples of resilience in the eyes of state and market withdrawal in deprived areas. Rather than focusing on deficiencies of the neighbourhoods, they have tried to find assets and maximise their capabilities, an example of 'Asset Based Community Development' (Gibson-Graham, 2006). For example through the incubation of different social organisations and creative goups, while offering their infrastructure to marginalised groups to use. However, as we have demonstrated this resulted in a change in social relations and some empowerment, although they have not yet realised their transformative potential.

Despite our findings, we must identify our limitations – firstly, the COVID-19 pandemic resulted in halt to almost all social centre activities of Romantso, and we therefore had to rely on word of mouth of those working in the space to imagine this function. In addition, a limited number of interviewees, almost all from the spaces, were used. Despite the limits, it opens new doors for future research. A deeper study with more actors among the assemblages than we have engaged with – namely, immigrants, locals from the area, officials from the municipality and others would be required to fully comprehend the dynamics of the assemblages in question. Finally, questions remain – how may CWS have a transformative impact in their local context, and what is the potential of translocal assemblages in this? How may they link with formal or informal institutions to enact this transformative potential? How may CWS in other contexts/assemblages engage in SI?

Acknowledgements This paper has received funding from the European Union's Horizon 2020 research and innovation programme under the Marie Skłodowska-Curie grant agreement No 955907, project CORAL (Exploring the impacts of collaborative workspaces in rural and peripheral areas in the EU).

References

Agustín, Ó. G., & Jørgensen, M. B. (2018). Solidarity and the 'refugee Crisis' in Europe. Springer.

Andersen, J., Delica, K., & Severin, M. (2013). From 'book container' to community Centre. In F. Moulaert, D. MacCallum, A. Mehmood, & A. Hamdouch (Eds.), *The international handbook on social innovation: Collective action, social learning and transdisciplinary research* (pp. 197–207). Edgar Elgar.

Avdikos, V., & Merkel, J. (2019). Supporting open, shared and collaborative workspaces and hubs: Recent transformations and policy implications. *Urban Research & Practice, 13*(3), 348–357. https://doi.org/10.1080/17535069.2019.1674501

Avdikos, V., & Pettas, D. (2021). The new topologies of collaborative workspace assemblages between the market and the commons. *Geoforum, 121*, 44–52. https://doi.org/10.1016/j.geoforum

Bandinelli, C., & Arvidsson, A. (2013). Brand yourself a Changemaker! *Journal of Macromarketing, 33*(1), 67–71. https://doi.org/10.1177/0276146712465186

Bandinelli, C., & Gandini, A. (2019). Hubs vs networks in the creative economy: Towards a "collaborative individualism". In R. Gill, A. C. Pratt, & T. E. Virani (Eds.), *Creative hubs in question: Place, space and work in the creative economy* (pp. 89–110). Springer.

Brown, J. (2017). Curating the "third place"? Coworking and the mediation of creativity. *Geoforum, 82*, 112–126. https://doi.org/10.1016/j.geoforum.2017.04.006

Chen, H. C., Knierim, A., & Bock, B. B. (2022). The emergence of social innovation in rural revitalisation practices: A comparative case study from Taiwan. *Journal of Rural Studies, 90*, 134–146. https://doi.org/10.1016/j.jrurstud.2022.02.003

Communitism. (2022, November 29). Retrieved from Communitism.space https://communitism.space/

DeLanda, M. (2006). *A new philosophy of society*. Bloomsbury.

DeLanda, M. (2016). *Assemblage theory*. Edinburgh University Press.

Deleuze, G., & Guattari, F. (1987). *A thousand plateaus: Capitalism and schizophrenia*. Bloomsbury.

Economou, C., Kaitelidou, D., Katsikas, D., Siskou, O., & Zafiropoulou, M. (2016). Impacts of the economic crisis on access to healthcare services in Greece with a focus on the vulnerable groups of the population. *Social Cohesion and Development, 9*(2), 99–115. https://doi.org/10.12681/scad.8880

Eurostat. (2022). *One in five people in the EU at risk of poverty or social exclusion*. Retrieved April 28, 2022, from https://ec.europa.eu/eurostat/web/products-eurostat-news/-/edn-20211015-1

Gandini, A., & Cossu, A. (2021). The third wave of coworking: "Neo-corporate" model versus "resilient" practice. *European Journal of Cultural Studies, 24*(2), 430–447. https://doi.org/10.1177/1367549419886060

García, M., Eizaguirre, S., & Pradel, M. (2015). Social innovation and creativity in cities: A socially inclusive governance approach in two peripheral spaces of Barcelona. *City, Culture and Society, 6*(4), 93–100. https://doi.org/10.1016/j.ccs.2015.07.001

Gibson-Graham, J. K. (2006). A Postcapitalist Politics. Minneapolis: University of Minnesota Press.

Hadjimichalis, C. (2014). Crisis and land dispossession in Greece as part of the global land fever. *City: Analysis of Urban Trends, Culture, Theory, Policy, Action, 18*(4–5), 502–508. https://doi.org/10.1080/13604813.2014.939470

Johns, T., & Gratton, L. (2013). The third wave of virtual work. *Harvard Business Review, 91*(1), 66–73. https://doi.org/10.3389/fpsyg.2013.00582

Malamidis, H. (2020). *Social movements and solidarity structures in crisis-ridden Greece*. Amsterdam University Press.

Marques, P., Morgan, K., & Richardson, R. (2018). Social innovation in question: The theoretical and practical implications of a contested concept. *Environment and Planning C: Politics and Space, 36*(3), 496–512. Available at: https://doi.org/10.1177/2399654417717986

McFarlane, C. (2009). Translocal assemblages: Space, power and social movements. *Geoforum, 40*(4), 561–567. https://doi.org/10.1016/j.geoforum.2009.05.003

Moulaert, F. (2013) The International Handbook on Social Innovation: Collective Action, Social Learning and Transdisciplinary Research. Edward Elgar Publishing.

Moulaert, F., & MacCallum, D. (2019). *Advanced introduction to social innovation*. Edward Elgar Publishing.

Moulaert, F., & Mehmood, A. (2020). Towards a social innovation (SI) based epistemology in local development analysis: lessons from twenty years of EU research. *European Planning Studies, 28*(3), 434–453. Available at: https://doi.org/10.1080/09654313.2019.1639401

Nail, T. (2017). What is an assemblage? *SubStance, 46*(1), 21–37. https://doi.org/10.1353/sub.2017.0001

Schmidt, S. (2019). In the making: Open creative labs as an emerging topic in economic geography? *Geography Compass, 13*(9), e12463. https://doi.org/10.1111/gec3.12463

Smith, T. S. J. (2020). "Stand back and watch us": Post-capitalist practices in the maker movement. *Environment and Planning A: Economy and Space, 52*(3), 593–610. https://doi.org/10.1177/0308518X19882731

Smith, A., & Dias, R. (2018). Making in Brazil: Can we make it work for social inclusion? *Journal of Peer Production, 12*(1), 43–59.

Teloni, D.-D., & Adam, S. (2018). Solidarity clinics and social work in the era of crisis in Greece. *International Social Work 61*(6), 794–808.

Vidaillet, B., & Bousalham, Y. (2020). Coworking spaces as places where economic diversity can be articulated: Towards a theory of syntopia. *Organization, 27*(1), 60–87. https://doi.org/10.1177/1350508418794003

Facilitating Social Economy in Taiwan: A Case Study of a University-Community Coworking Space in an Indigenous Community of Eastern Taiwan

Muyi Chou and Salizan Takisvilainan

Introduction

Economic development in post-war Taiwan followed the model of a developmental state, that is, state- or bureaucracy-led economic policies led to an export-oriented economy fully integrated into the world capitalist system. Moreover, Han Chinese culture has been the mainstream culture and language and capitalism has been the main approach of economic development for a long time. The Han Chinese culture and language and capitalist development have been reinforced by state policies, compulsory education, and mass media to the present day. Although minority rights were gradually given greater value after democratization, the policy ideology and mass media in general still tend to marginalize diverse ethnic groups and their cultures.

Regarding economic development, the capitalist economy plays a dominant role in policymaking and implementation at both the national and local level. Additionally, rural areas are gradually marginalized by the trend of rapid urbanization. Job opportunities are concentrated in cities, which discourages young people from living in rural areas. As a result, the devitalization of rural communities and their cultures has become a common challenge in Taiwan. Moreover, the capitalist market economy overly emphasizes competition and utility, which damages the solidarity of community and leads to negative effects on the environment, ecology, and social equality (Wright, 2010, p. 37). Profit-oriented production simplifies value as money. Profit maximization becomes the main principle to make an economic decision and leads to alternative considerations being neglected or deemed unrealistic. If

M. Chou (✉)
Bachelor Degree Program in Ocean Tourism Management, National Taiwan Ocean University, Keelung City, Taiwan

S. Takisvilainan
A Bunch of Millet Bunun Independent Publishing Studio, Hualien County, Taiwan

J. Merkel et al. (eds.), *Coworking Spaces*, https://doi.org/10.1007/978-3-031-42268-3_13

economic behavior separates itself from the local and cultural context, it is difficult to defend "value." Bringing culture back is key to creating value and preserving social and cultural diversity (Clammer, 2016, p. 10). When the mainstream economy is so dominant and threatens local autonomy, the rural area is viewed as a passive object. Is it possible to transform the rural area into an active subject and pursue alternative ways of development? Who can facilitate this change?

Regarding the actors involved in local civil society, more universities are embracing their social responsibility to contribute their knowledge and expertise to respond to the needs and challenges faced by local society. In Taiwan, several policy projects sponsored by the Ministry of Education and the National Science and Technology Council encourage universities to work with local communities. This trend has led many universities to become more committed to local affairs, social issues and developmental challenges by conducting university-community partnership research.

While university social responsibility was being encouraged by several government-sponsored pilot projects in Taiwan, the use of coworking spaces was becoming a global trend. Thus, various Taiwanese universities decided to construct coworking spaces as a means to build relations with local communities or civil society groups. Regarding literature on coworking spaces, one of the main debates focuses on whether coworking spaces reinforce the neoliberal culture of work or serve as an alternative to the mainstream economy (Gandini & Cossu, 2021; Vidaillet & Bousalham, 2020). However, further inquiries, such as what makes coworking spaces different from the capitalist paradigm and what makes coworkers feel a sense of community, deserve to be explored. In this research, we seek to provide some insights about these two inquiries through a case study in an indigenous community of Eastern Taiwan. Moreover, the coworking space in this case is not just an observable object, but rather an action taken by participants, including university-team members and community partners. We aim to illustrate how a university team builds partnerships with community partners and how they reframe "cultural practices" and "publishing" by assembling diverse economies.

This paper is structured as follows. Firstly, we review literature on social economy and diverse economies to establish the analytical framework. The academic debates on coworking spaces will be discussed to highlight what are the main concerns when coworking spaces are analyzed in the context of diverse economies. Regarding the method, this research is conducted using action research, that is, university teams serve not only as researchers but also participants working with local partners in carrying out projects. Subsequently, we introduce a case study of building a coworking space in an indigenous community using university-community partnership research and discuss how they reframe cultural practices and publishing through diverse economies. With this case, we show how the idea of social economy is translated into indigenous understanding and related to their cultural practices and how diverse participants can contribute to community development. Finally, we discuss the potential and the limitations of university-community initiatives. We argue that university-community partnership can facilitate the introduction of new forms of community-based social economy. Moreover, constructing a coworking space together can be a good start to build a relationship in which different participants feel they can make a meaningful contribution, which can serve as the basis to generate diverse economies.

Reframing Economy

In Taiwan, Han Chinese culture has been the mainstream culture and language, and capitalism has been the main approach of economic development. However, recently some rural areas have implemented community- or solidarity-based development alternatives to the competitive-based capitalist model. Huang (2022) takes Chishang, a famous rice township, as an example to illustrate how a township-level government and local society collaborate with each other to carry out alternative development by establishing Chishang Rice Mark and social redistribution. The township government turns the profits of Chishang Rice Mark into social welfare funds and redistribute them to farmers and township citizens to strengthen local identity and solidarity, which to some extent buffers them from the negative impact of neoliberalism on rural areas (Huang, 2022, pp. 224–231). Likewise, Chung and Shieh's (2021) findings show the development shift in three agricultural townships, namely, Dou-nan, Mei-nong and Chi-shang. Although driven by different mediating organizations, they all "establish institutional mix by flexibly combining three mechanisms for allocating resources, that is, market exchange, redistribution and reciprocity" rather than totally relying on market mechanisms (Chung & Shieh, 2021, p. 2). These innovative initiatives to some extent fix the side effects of market-oriented development and rebuild local solidarity. More importantly, they show rural development cannot be simplified as urbanization and that rural communities must be treated as autonomous entities rather than peripheral parts of cities. They have diverse potentials to create alternative economies, reframe development, and reshape their relations to cities and the international world (Klein & Pierre-Andre, 2013). In line with this, local development should be a "situated practice" to utilize local assets, respond to local needs, and take ethical consideration of culture, community and environment when making economic decisions (Amin, 2009, pp. 11–14; Gibson-Graham et al., 2013, p. xviii–xix; Borowiak, 2016, pp. 29–34). Theories and practices of social economy provide us alternative possibilities to envision development. Among them, "reframing" is a key to imagining an alternative economy, namely, viewing something familiar in new ways by giving new meanings (Gibson-Graham et al., 2013, p. 7–10). In practice, we wonder who can facilitate this reframing of an economy?

According to research on university-community partnerships in Quebec, universities located in different regions can play a crucial role to facilitate social economy initiatives (Bouchard, 2013). The Quebec experience is inspiring, especially for how it encourages the academic community to work with local communities and civic groups to develop conditions for initiating and sustaining the social economy. Recently, university social responsibility gradually became one of the main goals for the development of universities in Taiwan. At the same time, coworking spaces were becoming popular places for young people to initiate a small business or work as a freelancer. This new way of working was also conducted by combining the ideas of sharing economy, creative economy, knowledge economy, and social innovation while operating by non-hierarchical means. In this context, coworking spaces embedded the potential for generating social innovation, cross-disciplinary collaboration, sharing and reciprocity, and a sense of community.

However, some literature points out these ideal views on coworking spaces are not realistic and some coworking spaces reinforce the neoliberal culture or function as "another bubble in the knowledge economy" (Gandini, 2015, pp. 201–203; Grazian, 2020). Community in this sense is in name only or serves as "market intermediaries" for workers to build professional relationships and conduct practice-based projects without generating collective identity (Arvidsson, 2018). In contrast, another group of scholars provides evidence to show that coworking spaces can facilitate diverse economies and a sense of community (Vidaillet & Bousalham, 2020; Gandini & Cossu, 2021). Gandini and Cossu (2021) distinguish resilient coworking spaces from the mainstream, neo-corporate model. They define resilient coworking spaces as those that "embrace innovation and change, looking to set out practices of organization that adapt to the context they inhabit and exist in harmony with it" (Gandini & Cossu, 2021, p. 435). The core of their practices is to develop "actually communitarian" interaction, and it is viable when a coworking space is run by a worker co-op or connected to a sense of belonging to the local territory (Gandini & Cossu, 2021, p. 435; pp. 437–445). This task is similar to a community economy that seeks to reframe economy and build alternative economies by taking environmental and social impacts into consideration when making economic decisions (Gibson-Graham et al., 2013).

Regarding a sense of community, some literature focuses on the "affective turn" in explaining what builds a sense of community and how it controls or causes conflicts among participants (Waters-Lynch & Duff, 2019; Resch et al., 2021). Among these discussions, many research questions are derived from the "incoherent nature" (Gandini, 2015) of coworkers, such as working as an actual community versus "working alone together" (Spinuzzi, 2012), made up of both "precarious workers and new entrepreneurs" (Gandini, 2015, pp. 202–203). This contradictory coexistence is just like the social economy and has a similar challenge, that is, what should we do when achieving social purposes conflicts with economic performance? Or is it possible to co-exist? In Bouchard's (2013, p. 5) analysis, the social economy shows that the contradictory traits can viably coexist if economic activities or organizations fit five principles, including (1) the primary purpose is based on the members/community; (2) autonomy; (3) democratic decision-making; (4) profit-distribution based on members and work; and (5) operation based on participation, empowerment, and accountability. These principles are solidarity-based and crucial to balancing social purposes with the competitive and profit-oriented market economy. In line with this approach, we seek to situate coworking places in the context of social economy. Likewise, Vidaillet and Bousalham (2020, p. 60) propose the concept of "syntopia" to theorize "this type of place, whose main characteristic is that its alternative potential lies in enabling its users to articulate economic diversity." In this research, we intend to go a step further and explore what role culture can play in building a sense of community among diverse participants and facilitating diverse economic practices.

Moreover, reframing economy here not only refers to different ways of understanding economy but also includes different ways of practicing it. We find that Gibson-Graham, Cameron, and Healy's (2013, p. 13) framework of the diverse economy is helpful for us to identify diverse resources and exchanges, as Table 1 suggests. Some small projects in this case were designed by mixed uses of diverse economies. The logic of starting a social economy project is to reduce the reliance on currency for fulfilling participants' needs.

For practitioners who seek to effect change, the strategies of community development can be categorized into two main approaches. One is from an external perspective or needs-based community development strategy (Kretzmann & McKnight, 1993; Gibson-Graham, 2006, p. 145). This strategy is "problem-centered" and it diagnoses the lack of community and finds solutions through external assistance (Gibson-Graham, 2006, pp. 145–146). The other approach is from an internal perspective or asset-based community development (ABCD) strategy (Kretzmann & McKnight, 1993; Gibson-Graham, 2006, p. 145). Assets are interpreted in a broad sense here. They include not only various resources but also rights, capacities, capabilities, and access to assets in initiating change or creating economic opportunity (Nel, 2018, 37). ABCD is rights-based and usually carried out through cooperation, reciprocity, mutual help, and democratic participation, which cultivates community members as active citizens responsible for their future (Nel, 2018, pp. 36–37). Nel (2018, p. 46) compares ABCD-sensitized communities with non-ABCD-sensitized communities and finds that they both bring changes to communities, but the outcomes of trust-building are different. ABCD-sensitized communities build trust among community members as well as between the organization, facilitating development and community. However, "trust tends to exist more between the organization and community members and less among community members" in non-ABCD-sensitized communities (Nel, 2018, p. 48). Thus, a relationship-driven ABCD strategy is central to guide our action in the beginning stage. We think it is important to (re)build social ties and a sense of community as the basis of generating alternative economies.

Table 1 The diverse economy

Labor	Enterprise	Transactions	Property	Finance
Wage	Capitalist	Market	Private	Mainstream Market
Alternative paid	Alternative capitalist	Alternative market	Alternative private	Alternative market
Unpaid	Noncapitalist	Nonmarket	Open access	Nonmarket

Source: Gibson-Graham et al. (2013, p. 13)

Methods

The empirical evidence of this article originates from part of the Humanity Innovation and Social Practice (HISP) Project conducted by the HISP team of National Dong Hwa University (the NDHU team hereafter) in the period 2015–2018. The first author joined the HISP project in August 2015 and served as a postdoctoral fellow in the HISP coordination office which was the bridge between the sponsor, the National Science and Technology Council, and the grantees and the university teams. She was responsible for overseeing the processes and challenges of conducting this project, which allowed her to participate in university teams' work and have fruitful discussions with team members and community partners. At that time, there were four university-teams implementing HISP projects in different places in Taiwan, including at National Dong Hwa University (NDHU). This case and site were conducted by the NDHU team and their HISP project ended in June 2018. The second author is a Bunun poet and the founder of A Bunch of Millet Bunun Independent Publishing Studio (Millet Studio hereafter), which is the coworking space this article focuses on. Salizan felt that where he lived and received education was dominated by Han Chinese culture, which made him gradually distance himself from Bunun essentials. He was afraid of the loss of Bunun language. Thus, he expressed his love and nostalgia for his tribe by writing poems. Later, he had the idea of publishing them to make Bunun books more accessible to Bunun people and to invoke his concern for Bunun culture and language. In 2015, he went back to Nakahila and prepared to initiate Millet Studio, which provided a chance for the NDHU team to work with him in "Nakahila." They worked together from 2015 to 2018.

When the first author joined the HISP project, it was at the beginning stage of Millet Studio. She visited the site every 2 months and stayed 2–3 days for each visit. Through these regular visits, she could join a variety of activities organized by Millet Studio, including fieldwork, crucial events, cultural activities, workshops, book launch parties, etc. The core members of the Millet Studio project include Salizan and the research assistant of the NDHU team. Other NDHU team members including professors and postdoctoral researchers in different disciplines also participated in this project by contributing their knowledge, participating in field studies and community-engaged courses, and supervising students' independent studies. Salizan, the research assistant and several postdoctoral researchers are all alumni of NDHU. NDHU was the catalyst that gathered scholars, students, and community partners at the beginning stage. Salizan was the key liaison and served as the bridge between the NDHU team and Nakahila.

Regarding data collection, besides working with the projects of Millet Studio, it also includes the first author's field notes from participant observation and semi-structured interviews with participants about their experiences, thoughts, and feelings while working in Millet Studio. The second author established and runs Millet Studio and plays an important role in this university-community collaborative project. He interviews the Bunun elderly to record the language, history, stories, traditional ballads, and cultural practices and collaborates with the NDHU team to transform these materials into a variety of publications and activities. He provides other participants with knowledge about Bunun culture and shares his poems and inspiration with us. Such collaboration also creates new ways of knowledge production and knowledge transfer.

This project was conducted using action research, meaning that university team members were not only researchers but also participants working with community partners. Moreover, researchers had to be culturally sensitive to capture the insights in line with local context and understanding (Lake & Wendland, 2018, p. 17). Thus, how to localize the theoretical concepts and methods is decisive when conducting action research. We think culture serves as an important role in shaping a sense of community, localizing the concept, and operationalizing the project. According to Cameron and Gibson (2020, p. 511), action research for diverse economies usually adopts three main strategies, including "activating new languages, shifting subjectivities, and collective action." In other words, using different ways or words to describe a phenomenon can shape how we perceive and perform it, which is helpful to reframe questions and challenges. Moreover, everyone inhabits multiple roles and identities (subject positions), that is, the individual subject is not a singular whole but rather a diverse assemblage with the potential to carry out diverse relations to economic practices. Combining the new language of economy and the decentered subject may encourage collective actions to practice diverse economies. The framework of diverse economies and these three strategies serve as an analytical framework for our case study. In addition, we will show how we localize and enrich the ideas of social economy with Bunun culture and carry out a series of ABCD projects.

A Case Study of A Bunch of Millet Bunun Independent Publishing Studio in Nakahila

Nakahila is an indigenous community of Taiping village in southern Hualien in eastern Taiwan. There are about 40 households living there. More than 90% of the residents are Bunun people. Local organizations in the area include a community development association, a Catholic church and a Christian church. The local economy relies on agriculture, which is mainly maintained by the aging residents. Most young people have no choice but to move to cities for job opportunities, which means Nakahila is confronted by demographic decline and the loss of Bunun culture and language. Thus, the preservation of Bunun culture and language is the main task for this university-community collaborative project.

Building a University-Community Partnership by Creating a Multi-functional Coworking Space

At the beginning, the NDHU team worked with the Community Development Association (CDA) to plan a suitable partnership with the community. However, the project director lost the election and was replaced by his competitor who was unwilling to continue the cooperation. Thus, the NDHU team had to find other community partners to work with. Around the same time, a young Bunun poet, Salizan, returned to Nakahila and prepared to start a publishing house for Bunun cultural preservation. Salizan graduated from the NDHU and knew the NDHU's assistant during college. This alumni connection brought Salizan together with the NDHU team, and he soon became the project's key local coordinator. Preserving Bunun culture was their main purpose. The building where Millet Studio was located is a house in Nakahila that was owned by a Bunun elder. He rented it to Salizan for 5 years in exchange for rebuilding the broken roof (Hsieh, 2021, p. 199). This was where Millet Studio began.

Regarding the university-community collaboration, Salizan possessed the knowledge and ability to investigate and collect Bunun culture and traditional techniques, while the NDHU team had various methods to display or represent them. The NDHU team invited scholars from different disciplines, and they applied various methods (e.g., open street map, films, publications, paintings, community-engaged courses, etc.) to record Bunun culture and give feedback to the community. Culture was the foundation for university-community collaboration in Nakahila, as culture-based activities are easy for residents to join regardless of their political or interest groups. Moreover, doing something together was also a process for participants to find their roles and ways to work with each other. This coworking space provided opportunities for participants to experience different things, explore themselves, be open to different cultures and working styles, and enjoy life, which was similar to the space that Vidaillet and Bousalham (2020) described as syntopia.

The establishment of Millet Studio was the outcome of collective efforts, including community members, Bunun intellectuals, the NDHU team, and so on. The university-community partnership was built by designing community-engaged courses based on community partners' needs. In this stage, the NDHU team and community partners constructed relationships through some small-scale coworking projects and aggregated trust accordingly. For example, professors and students from the department of art and creative industries helped with the renovation, created wall paintings for the space, and gave advice for the creative design of small products to promote Bunun culture (Hsieh, 2021, p. 202). One of these wall paintings was based on Salizan's poem and hung in the front yard to represent the main purpose of Millet Studio:

> The hunter's footprints
> do not (intend to) lead me to climb a mountain,
> but rather to know the ancestors' paths.
> Source: Salizan's (Takisvilainan, 2013, p. 28) poem in *The Lights of the Tribe*.

These activities were not entirely planned in advance. Instead, many ideas were created during informal chats or in the process of collaboration between different participants. Needs and wishes were collected and discussed during fieldwork or chats with the elderly. Then the NDHU team and Salizan sought to transform these needs and wishes into workable plans. Thus, these spontaneous ideas were not just words but had a chance to become realized.

Chou et al.'s (2018) "bottom-up, local, open, and reassembling (BLOR) actor network" is useful to reconsider "local." It is not a closed network, but rather is inclusive and open to public participation and linked to the wider society. In line with this concept, Salizan and the NDHU team mobilized several resources from within and outside. Firstly, they renovated an unused building together. By working together, it created a tie and a sense of "us" among diverse participants. This was realized by several community-engaged courses conducted by the NDHU College of Art and the NDHU College of Indigenous Studies. These courses were designed by community partners, the NDHU team, and the professors to ensure practical projects were consistent with community needs. During the process, students pitched some ideas, and if there was one that community partners also wanted to try, the NDHU team and students worked with community partners to realize it. Those who proposed ideas also carried them out. The secret is to put the person in the right position to exert his/her capability. Salizan and the NDHU assistant were well-acquainted with local people in Nakahila and what they were good at. This was the base to seek people who can work with.

Secondly, in order to let more people know about Millet Studio and use this place, Salizan came up with the idea of having a community library. He hoped local children could access indigenous books easily and learn about indigenous culture. The interior decoration of Millet Studio followed this idea. Moreover, Salizan and the NDHU team invited people to donate books related to indigenous studies and culture to Millet Studio. Through this book donation activity, many scholars and cultural intellectuals around Taiwan learned about Millet Studio and contributed their collections to it. In order to encourage residents to use this place, they organized extracurricular activities for children during summer vacations. Children came to play and read with student volunteers' company and guidance. Creating opportunities for gathering was the key in this stage.

Ultimately, the renovation work was completed in October 2015 and celebrated with a Bunun traditional ceremony known as "pig-slaughtering and pork-sharing feast" to share the happiness of the inauguration with community people and thank the NDHU team and students for their contribution. Every household received "a bag of pork" which means "to share it with you" (Hsieh, 2021, p. 203).

Cultivating cultural sensitivity was the key for outsiders who wanted to work with an indigenous community. Inspired by the Bunun pig-slaughtering and pork-sharing feast to show the host's thanks, the NDHU team revised their mode of cooperation and sought different means to work with community partners. Experiential knowledge embedded in their daily life could enrich the methods and prospects of university-community collaboration. The cultural learning course was designed accordingly.

The cultural learning course was different from routine courses. It had flexibility in schedule and in content. Community partners were instructors and led students to understand how their culture was derived from daily life. For example, millet is a staple food in Bunun culture. Many traditional Bunun ceremonies accompany the cultivation of millet, such as the ceremony of reclaiming land, throwing rocks, seeding, collecting tools, repelling birds, harvest, ear-shooting, and hunting gun. These ceremonies also take place along with the four seasons, which shows the Bunun's humble attitude and respect for the spirits and Nature. Nowadays, only a few Bunun elders remain to grow millet because it is quite labor intensive. Through this course, students participated in planting millet with Bunun elders and learned the skills and cultural practices from them. They also helped to record the process in words, photos, and films as supplements to the elders' customary way of delivering knowledge by telling and doing.

To sum up, building a coworking space created an opportunity for the NDHU team and community partners to work together. Learning Bunun culture by doing made this a place not only for learning but also gathering.

Reframing Publishing for Culture Preservation

In this case, we framed the local according to the ABCD strategy, which emphasizes various local assets, including "personal, physical, financial, natural, political, social, spiritual, and cultural assets" (Nel, 2018, p. 37). Bunun culture was the key element of local assets here. Salizan continued interviewing the elders and recording Bunun history, legends, cultural practices, and language. These precious materials were collected and edited to become books which were conducive to the preservation of Bunun culture. The NDHU team and students were responsible for promotion activities. They discussed proposals with Salizan and then carried them out together. Students contributed to typesetting, cover design, postcard making, and marketing. The related artwork was designed using Bunun cultural elements. Students tried to visualize their understanding of Bunun culture and turn it into limited-edition products.

Besides university partners, a group of young people who worked and lived in Southern Hualien were also very supportive of Millet Studio. One of the young partners helped lead a practical course in Nakahila and then stayed in a nearby part of Southern Hualien to work. She initiated "let's have a dinner party together" at Millet Studio every 1 or 2 months. People who joined the dinner were welcome to bring friends to the casual gathering. Every participant brought a dish or drinks to share with others. There were some new friends each time. Participants shared their hobbies and interests with others. It gradually became a small talent pool for future cooperation (Hsieh, 2021, pp. 204–206). They not only helped to edit, proofread, and design the books but also threw book launch parties. Their participation enriched Millet Studio with lots of joy and creativity.

Diverse participants joined Millet Studio through informal gatherings or formal courses, research projects, and cultural activities. They shared interesting ideas and carried them out together. This was the driving force for participants to join and contribute to Millet Studio. This way of running Millet Studio was very different from formal organizations. It was not profit-oriented. Instead, its work and activities were "culture-based" and "community-based," which cultivated a humane and mutually supportive working culture. The flexible way of working allowed participants enough space to explore how to situate themselves in Millet Studio, for example, by painting, singing, cooking, decorating, or writing. The particularities of Nakahila were based on Bunun culture, and participants were nourished by it. In return, they used diverse ways to enrich Millet Studio.

Through these collective efforts, Millet Studio published seven books in the Bunun language while cooperating with the NDHU team from 2015 to 2018 (Hsieh, 2021, p. 208). Two of them are portable dictionaries of two Bunun dialects which are very important to the preservation of Bunun language diversity. In order to encourage more Bunun teenagers and children to learn Bunun language, Millet Studio also builds on-line dictionaries with pronunciation guides and some videos to introduce Bunun culture. With Bunun book publishing, Millet Studio became a place to learn Bunun culture. More importantly, some Bunun writers were willing to let Millet Studio publish their work. It implies that many Bunun people think it is important to preserve culture and language and many of them have done it in their own ways for a long time. However, their work rarely has a chance to be published by the mainstream, market-oriented press. Thus, Millet Studio serves as an amplifier to help more people know Bunun culture and encourages more Bunun people to write their stories. More importantly, publishing Bunun books to preserve the diversity of Bunun language serves as an alternative mechanism to challenge the current ethnic group policy that distributes resources based on population in a way that is unfair to the minority subgroups of an ethnic group. The majority subgroup receives more resources than the others, which leads to the language and culture of minority subgroups becoming even more marginalized. In 2016, Millet Studio was registered as a studio. It was an expedient choice because the process of registering as a publisher was easier than that for registering as a nonprofit. Being registered as a formal organization is important for Millet Studio to aggregate collective work.

In sum, the aim of this university-community collaborative project in Nakahila was to explore local particularities and utilize these assets to pursue local-based development. Some small projects were based on community needs, i.e., the renovation of the space and extracurricular activities. The others were related to ABCD which activated community resources and capacities and expanded the network of people promoting Bunun culture, i.e., publishing books and book launch activities. Even the community needs in this case were also for cultural purposes.

During the university-community collaboration, we found the social economy and Bunun culture had something in common. For example, the reciprocity, cooperation, community orientation, and diverse economy based on nonmonetary exchange in a community economy are similar to the "labor exchange" employed in Bunun culture when building a traditional house in the past. This was the way that

Bunun people helped each other and Bunun men learned the knowledge and the skills about building a traditional house by helping others to do it (Salizan, 2019, p. 165). Moreover, in the past, only one family was entitled to weave. Other Bunun people "bartered" in exchange for woven items, such as bags, headbands, sieves, and so on (Salizan, 2019, p. 24). Furthermore, in Bunun hunting culture, the virtues of collective well-being, altruism, solidarity, cooperation, and sharing were crucial and practiced by hunters (Takisvilainan, 2017, pp. 182–183). The hunter hunted for collective needs and had to share the prey with other Bunun people. As a Bunun proverb said, "be generous to others, otherwise God will not give you generous things anymore," which implies if you share things with others, your hands become empty and then able to grasp a new harvest (Takisvilainan, 2017, p. 185). These virtues are embedded in Bunun culture, which also inspires us to sustain Millet Studio by assembling a diverse economy.

Sustaining the Millet Studio Through a Diverse Economy

Table 2 presents the different aspects of the diverse economy of Millet Studio. Most projects relied on collective efforts and diverse resources internally and externally. Take book publishing, for example. It was unlikely such a small publishing studio could survive, especially if it only published Bunun books since the market was very limited from the capitalist perspective. However, it was possible from the perspective of social economy because Millet Studio worked with a university team, and they assembled diverse economies creatively. Among them, labor exchange was crucial for binding participants by working together.

In this case, Millet Studio relied on "labor exchange," "volunteers," "university partners," and "friendship" to minimize the monetary exchange when running the projects. For students, their designs had a chance to be published. For the community elders, their wisdom and knowledge could be treasured and inherited by different forms and means. For Millet Studio, each publication brought some new friends, and Bunun knowledge and culture were gradually accumulated. Some books were gifts to community members who joined the book launch parties. The Millet Studio reframed the publishing from a business into a cultural platform to promote and preserve Bunun culture and language. The books were gifts to deliver Bunun culture and a joy to share with participants. Although it was not financially stable, Millet Studio tried diverse ways to link and mobilize internal and external resources. Most participants had full-time jobs and spent their leisure time working with Millet Studio. For professional support such as editing, translating, and art design, they were paid on a piecework basis. The flexible and open-minded attitude of Millet Studio made participants joyful and willing to work with it. The informal gathering was equally important. When working together at Millet Studio, participants usually cooked and ate together, and they sometimes had a barbeque after finishing a difficult task. Casual gatherings, after parties (following a big event), and small talks made their interaction more like that of friends than colleagues.

Table 2 The diverse economy in Millet Studio

Labor	Enterprise	Transactions	Property	Finance
Paid labor Staff at the Univ.	Capitalist –	Market Printing	Private –	Mainstream Market finance –
Alternative paid Self-employed poet Home-based graphic designer Part-time staff for Bunun book projects Labor exchange Culture teaching and learning	Alternative capitalist –	Alternative market Reciprocal exchange Bunun-book publishing Bartering	Alternative private The house and place of Millet Studio (owned by neighbor but used for Millet Studio)	Alternative market Governmental grants
Unpaid Volunteers Work (courses and projects) with the Univ. Support from friends and community partners	Noncapitalist Bunun-based publishing Studio, "Millet studio" Self-employed business	Nonmarket Book donation Gifting Borrowing Promoting Bunun culture by book launch parties	Open access On-line Bunun dictionary and videos that introduce Bunun culture	Nonmarket Family lending

Conclusion

Millet Studio published Bunun books, which gave a marginalized language and culture a chance to be preserved and introduced to the public. The publishing was reframed as a means of preserving Bunun culture by facilitating cross-disciplinary and cross-community collaborations. Although Millet Studio applied for several governmental grants to run the projects and publish books, it redistributed these governmental grants to make up for the insufficiency of the current ethnic group policy. The working atmosphere of Millet Studio was characterized by a flexible and open-minded attitude, and each project recruited new members, which shows it functioned as an open network. One of the workable mechanisms to keep it open was through providing practical courses under a university-community partnership. Unlike Nel's (2018) analysis that took a community as a whole, the community should be viewed as a heterogenous constitution here. Community members had different attitudes toward the community-university partnership and maintained different kinds of relations with the university team. In this case, both ABCD and needs-based strategies were conducted. For the core participants, their work followed the logic of the ABCD strategy and a process of empowerment (Chou, 2018).

For community members who were uncertain, a needs-based strategy and achievement sharing were useful to better publicize community-university collaborative projects. The relations between them remained friendly which made it easier for community members to later become core participants. Thus, stepping forward from a needs-based strategy was necessary, and it was exactly what the university-community partnership did here. In line with Cameron and Gibson's (2020) notions of reframing and the decentered subject in facilitating diverse economies, university-community partnership in this case reframed needs and challenges and activates participants' talents and potential by utilizing participants in multiple roles and putting each person in the right position to exert his/her capabilities. This mutual trust and understanding were key to bind participants and help them work in solidarity.

Nevertheless, this kind of collaboration was limited due to the semester schedule and students' workload. It was workable at the beginning stage but not suitable for issue-based projects. Moreover, the organizational form of Millet Studio needed to be reconsidered. It registered as a studio but functioned as a nonprofit organization. There is no compatible status for socially innovative organizations in Taiwan's current laws. Finally, regarding the university-community partnership, the different working culture and pace between universities and communities require lots of communication to reduce the inevitable gaps.

In conclusion, the contribution of this paper is to provide an alternative mechanism to run a coworking space based on the indigenous culture and social economy embedded in it. The coworking space here is viewed as a series of actions through which diverse participants exchange different needs and supplies in order to promote cultural values. Moreover, this case displayed how a sense of community is built by different participants. Millet Studio showed its inclusion and resilience to work with different people and its capacity to mobilize resources internally and externally for the purpose of creating more interactions between Bunun culture and people. The virtues of sharing, altruism, and collective well-being and the traditional practices of bartering and labor exchange embedded in Bunun culture resonated with ideas of social economy and inspired participants to practice and localize diverse economies. Instead of the profits, it was the process of working together that tied participants together and made them feel a sense of meaning and belonging in this place. What made participants feel meaningful also served as the potential to generate diverse economies.

References

Amin, A. (2009). Locating the social economy. In A. Amin (Ed.), *The social economy international perspectives on economic solidarity* (pp. 3–21). Zed Books.

Arvidsson, A. (2018). Value and virtue in the sharing economy. *The Sociological Review, 66*(2), 289–301. https://doi.org/10.1177/0038026118758531

Borowiak, C. (2016). Mapping social and solidarity economy: The local and translocal evolution of a concept. In N. Pun, B. H.-B. Ku, H. Yan, & A. Koo (Eds.), *Social economy in China and the world* (pp. 17–40). Routledge.

Bouchard, M. J. (2013). Introduction: The social economy in Quebec: A laboratory of social innovation. In M. J. Bouchard (Ed.), *Innovation and the social economy: The Quebec experience* (pp. 3–24). University of Toronto Press.

Cameron, J., & Gibson, K. (2020). Action research for diverse economies. In J. K. Gibson-Graham & K. Dombroski (Eds.), *The handbook of diverse economies* (pp. 511–519) Edgar Elgar.

Chou, M. (2018). Social practice from a capabilities approach perspective. In R.-M. Tsai (Ed.), *New praxis and local societies* (pp. 315–350). The Office of the Humanity Innovation and Social Practice Project.

Chou, M., Xiong, H.-L., & Chen, D.-S. (2018). Social innovation networks in practice: A case study of the humanity innovation and social practice project. *The Taiwanese Political Science Review, 22*(2), 147–120. https://doi.org/10.6683/TPSR.201812_22(2).0004

Chung, Y.-T., & Shieh, G.-S. (2021). The dynamics of agriculture: Market exchange, redistribution and reciprocity in rural Taiwan. *Taiwanese Sociology, 41*, 1–50.

Clammer, J. (2016). Social economies and economic anthropology: Challenging conventional economic thinking and practice. In N. Pun, B. H.-B. Ku, H. Yan, & A. Koo (Eds.), *Social economy in China and the world* (pp. 3–16). Routledge.

Gandini, A. (2015). The rise of coworking spaces: A literature review. *Ephemera: Theory & Politics in Organization, 15*(1), 193–205.

Gandini, A., & Cossu, A. (2021). The third wave of coworking: 'Neo-corporate' model versus 'resilient' practice. *European Journal of Cultural Studies, 24*(2), 430–447. https://doi.org/10.1177/2F1367549419886060

Gibson-Graham, J. K. (2006). *A postcapitalist politics*. University of Minnesota Press.

Gibson-Graham, J. K., Cameron, J., & Healy, S. (2013). *Take back the economy: An ethical guide for transforming our communities*. University of Minnesota Press.

Grazian, D. (2020). Thank God it's Monday: Manhattan coworking spaces in the new economy. *Theory and Society, 45*, 991–1019. https://doi.org/10.1007/s11186-019-09360-6

Hsieh, P.-Y. (2021). Becoming a bunch of millet Bunun independent publishing studio. In K.-H. Hsieh (Ed.), *Social practices in Taiwan* (pp. 196–209). The Office of the Humanity Innovation and Social Practice Project. Retrieved May 15, 2022, from https://www.hisp.ntu.edu.tw/ebook/8

Huang, S.-W. (2022). *Making Chihshang*. Rive Gauche Publishing House.

Klein, J.-L., & Tremblay, P.-A. (2013). The social economy: A springboard for local development project? In M. J. Bouchard (Ed.), *Innovation and the social economy: The Quebec experience* (pp. 229–254). University of Toronto Press.

Kretzmann, J., & McKnight, J. (1993). *Building communities from the inside out: A path toward finding and mobilizing a community's assets*. Asset-Based Community Development Institute, Institute for Policy Research, Northwestern University.

Lake, D., & Wendland, J. (2018). Practical, epistemological, and ethical challenges of participatory action research: A cross-disciplinary review of the literature. *Journal of Higher Education Outreach and Engagement, 22*(3), 11–42.

Nel, H. (2018). A comparison between the asset-oriented and needs-based community development approaches in terms of system changes. *Practice, 30*(1), 33–52. https://doi.org/10.1080/09503153.2017.1360474

Resch, B, Hoyer, P., & Steyaert, C. (2021). Affective control in new collaborative work: Communal fantasies of purpose, growth and belonging. *Organization studies, 42*(5), 787–809. https://doi.org/10.1177/2F0170840620941616

Salizan. (2019). *Carrying mountains by the headband: Stories of mountain guides and rangers*. Chien Hsing Publishing.

Spinuzzi, C. (2012). Working alone together: Coworking as emergent collaborative activity. *Journal of Business and Technical Communication, 26*(4), 399–441. https://doi.org/10.1177/1050651912444070

Takisvilainan, S. (2013). *The lights of the tribe*. Mountain and Sea Culture Publishing.

Takisvilainan, S. (2017). *Becoming a Bunun: Bunun childhood and education*. A Bunch of Millet Bunun Independent Publishing Studio. 【The Humanity Innovation and Social Practice Project. Retrieved May 15, 2022, from https://www.hisp.ntu.edu.tw/english】 This is another reference and should be separated from this one.
Vidaillet, B., & Bousalham, Y. (2020). Coworking spaces as places where economic diversity can be articulated: Towards a theory of syntopia. *Organization, 27*(1), 60–87. https://doi.org/10.1177/2F1350508418794003
Waters-Lynch, J., & Duff, C. (2019). The affective commons of coworking. *Human Relations, 74*, 383–404. https://doi.org/10.1177/2F0018726719894633
Wright, E. O. (2010). *Envisioning real Utopias*. Verso.

Spaces of Collaboration-Spaces of Possibility: How a Local Hub Creates New Possibilities in a Greek Peripheral Town

Danai Liodaki

Introduction

This chapter draws from literature on coworking and collaborative workspaces (hereby CWS) and analyses the case of the local hub, "Messolonghi by locals", placed in the Greek peripheral city of Messolonghi. Firstly, main arguments regarding the transformative potentials of CWS are presented. As stated in the next section, the current literature on CWS focuses on the categorizations of the different initiatives and questions whether they can be socially beneficial or not. Nevertheless, it is worth wondering in which ways the CWSs can – or at least attempt to – create social change in the places they operate. This question is being approached by empirical research in the local cultural hub "Messolonghi by locals". In the following sections, I aim to unravel in which ways this hub attempts to create new possibilities for the community of Messolonghi and beyond. How do the people involved in this initiative open the place to new practices and reform the local identity to create social change in their region?

This question is approached through empirical research in the local hub. Firstly, I present some information about the city and the organization's activities and give an insight of the founders' main motives and objectives. In order to have a better understanding of the hub's practices and objectives, digital material from the organization's website was used and is presented in this chapter. The initiative's practices were observed through their social media and digital presence since the day of the organization founding. Informal talks with the founding members were also conducted in several times since 2018. In 2022, semi-structured interviews with founding and core members were conducted. During these interviews the members of the initiative described the history of their organization, their practices, their

D. Liodaki (✉)
Leibniz Institute for Regional Geography and Leipzig University, Leipzig, Germany
e-mail: d_liodaki@leibniz-ifl.de

J. Merkel et al. (eds.), *Coworking Spaces*, https://doi.org/10.1007/978-3-031-42268-3_14

operation, their connection with other local actors as well as their personal and collective motives for their participation in the initiative. The data collected by the digital observation, the informal talks and the semi-structured interviews were used for the analysis and the framing of the organization's practices in the three categories presented in this chapter. The focus is on the members' approach and objectives, on their strategy and motives and the ways in which they are attempting to promote social transformation in their region. If and to which extent they are succeeding remains an open question for further research.

Spaces of Collaboration-Spaces of Possibility

I argue that CWS can become spaces of possibility and can create new potentials for social, economic and political transformation in the regions where they operate. To support this approach, below I present the current situation concerning CWS and their social role, the critique they have been under and the line of literature that highlights their transformative potentials.

The phenomenon of coworking was firstly observed and named – or advertised – as such in 2005 in San Francisco, although similar coworking practices were performed before (Brinks, 2022, p. 4). Coworking has been considered as a "third way" of working, in-between of working in a traditional environment and specific colleagues and working alone as a freelancer – mainly in the cultural industry. So, the term tries to depict the in-between practices of working individually in a shared space (Gandini, 2015, p. 195). After 2005 the practice of coworking and the emergence of coworking spaces have grown broader and broader mainly due to the digitalization of the cultural and knowledge sector (Schmidt et al., 2014, p. 234), "the rise of the creative economy and the sharing economy paradigms" (Avdikos & Iliopoulou, 2019, p. 111), the financial crisis that led many freelancers to look for cheaper alternatives in their working arrangements and others. In the years that followed, people using coworking spaces were radically raised, "accounting for 1.74 million workers, four-fifth of whom in Europe and North America" (Deskmag, 2018 in Vidaillet & Bousalham, 2020, p. 62), while projections show that more than 5 million users will be placed in collaborative workspaces in 2022 (Avdikos & Pettas, 2021, p. 44). In parallel, researchers have broadened the definition of coworking and now speak about collaborative workspaces (CWS) that include "coworking spaces, hubs, hackerspaces, business incubators, accelerators and the like" (Avdikos & Pettas, 2021, p. 44).

The phenomenon of CWS has been widely observed and analysed from a range of disciplines such as spatial sciences (e.g. Avdikos & Kalogeresis, 2017), management (e.g. Capdevila, 2015), policy (e.g. Avdikos & Merkel, 2019) and others. Nevertheless, coworking and CWS are not yet fully defined, while there are scholars that have identified major inconsistencies and differences between different CWS and coworking practices. Specifically, Spinuzzi has argued that after describing "so many contradictions", one "might suspect that coworking does not even describe a coherent phenomenon" (Spinuzzi, 2012, p. 428). In that sense, the case that this chapter focuses on, the local hub "Messolonghi by locals", shares some common elements with traditional coworking spaces but also hosts operations of a community centre, a creative hub, or community activism.

There is a lot of research emphasizing in the ways CWS can be beneficial for coworkers, broader communities or the progress in different sectors by improving the workers' well-being (Mariotti and Akhavan, 2020), fostering innovation (Capdevila, 2015) or contributing to a region's development (Capdevila, 2013). As Gandini argues, coworking was from its start about community and a communal approach to work, a practice that aimed to cultivate communitarian relations among the people involved, and not a phenomenon oriented to business (Gandini, 2015, p. 196). Some of the collaborative workspaces that emerged during the last decades have been "both innovative and alternative" and have been based around progressive values such "autonomy in work, empowerment, social experimentation, collaboration and accountability" (Vidaillet & Bousalham, 2020, p. 62). The last years there has also been a growing literature about CWS that can act as commons and host processes of care and commoning (e.g. Avdikos & Iliopoulou, 2019, De Peuter et al., 2017, Merkel, 2015, Waters-Lynch & Duff, 2019, Avdikos & Pettas, 2021). Some emphasize the internal relations of a CWS, the everyday activity in them, the immaterial practices they foster, the curatorial practices implemented to create a sense of community and a caring atmosphere for people involved (Waters-Lynch & Duff, 2019, pp. 2–4). Other scholars focus more on the transformative perspectives in CWS and their "capacity to challenge the neoliberal politics of individualization" (Merkel, 2019, p. 531) or even their contribution in fostering diverse economic activities that challenge the capitalocentric approach of economy – as Gibson-Graham (2006) put it (Vidaillet & Bousalham, 2020).

Nevertheless, there is also a critical approach to CWS and their beneficial potential. Gandini argues that this approach of coworking spaces as spaces where community can emerge and where competition is not activated "remains a challenging issue" (Gandini, 2015, p. 198). As he explains coworking was already from 2015 "being increasingly used for branding, marketing and business purposes" and it has created a "bubble"; it was transformed into a "buzzword" with too high expectations in relation to what it could achieve (Gandini, 2015, pp. 193–194). Other scholars have claimed that coworking even though it started as a "self-proclaimed grassroots movement" has evolved into a very "neo-corporate" trajectory that promotes "flexible work, incarnated by global franchise giants such as WeWork" (Gandini & Cossu, 2021, p. 431), and community is now only an "empty signifier in coworking parlance" (de Peuter et al., 2017, p. 9). Even in the personal level of coworkers' well-being, there have been critical points and scholars that argue that although coworking can have beneficial aspects for the affective state of people involved, there are also negative effects that can be observed, such as feelings of failure and others (Lee, 2019, p. 71). Especially for creative hubs, scholars have connected their emergence with gentrification processes (Dinardi, 2019) and have pointed out that the relevant research focuses on the "real estate dimension" (Pratt, 2021, p. 6), instead of the users and their experience. Gandini also highlights the difficulties and limitations for CWS to act as political catalysts, since precarious freelance work is implemented in a "context made of limited unionisation and politicisation, and very little self-reflexivity" (Gandini, 2015, p. 203).

Building on those contradictions, between approaching CWS as progressive, alternative, socially beneficial places able to foster political and social change and approaching them as a neoliberal, or neo-corporate buzzword, many scholars have worked on contrasting categorizations of CWS. In that sense Avdikos and Pettas (2021, p. 47) explore the position of CWS "between the commons and the capitalist market", Avdikos and Iliopoulou (2019) explore the differences between "community-led" and "entrepreneurial-led" CWS, Gandini and Cossu (2021) approach coworking spaces as "neo-corporate" or on the other hand "resilient" and so on. Even though most scholars admit that there are many CWS aligned with the neoliberal spirit of precarity and antagonism, there are others arguing that CWS could still be a "platform of collective action" contributing to a postcapitalist perspective (de Peuter et al., 2017, p. 3). In parallel, there are lines of thought that go beyond this dual way of perceiving CWS and focus more on their openness and their ability to create new possibilities. For example, Vidaillet and Bousalham (2020) conceptualize coworking spaces as "syntopias" in the sense that "juxtapose heterogeneous and potentially contradictory aspects that come with the economic diversity inherent to a post capitalocentric economy" (Vidaillet & Bousalham, 2020, p. 80). Merkel (2019) specifically proposes to go beyond a logic of "opposition versus co-optation" (Elwood, 2006 in Merkel, 2019, p. 531) and regard coworking spaces as "spaces of possibility" following Massey's conceptualizations (Massey, 1994, 2005). This chapter aims to focus on such perceptions of CWS as spaces that could host the coexistence of multiple narratives, practices and encounters and open new possibilities for the communities and regions around them.

The Case of "Messolonghi by Locals"

The case under research is "Messolonghi by Locals", a local hub formed as a non-profit organization, founded in 2018 in the city of Messolonghi. Messolonghi is a city in the Region of Western Greece. The Region of Western Greece is one of the poorest regions of Greece and is considered as relatively underdeveloped or even left behind. The city of Messolonghi follows the same pattern. It is a small city, with approximately 35,000 residents. The population in the region and the city is declining and ageing the last decades. The population of Messolonghi that is over 60 years old is more than 25% of the overall population. These issues are also mirrored in the percentage of the economically inactive population (people that do not work, minors, retired people, people that live from the support of others) that has risen to almost 65% of the overall population. Generally, the city follows the pattern of the region also in the percentages of people working in the service sector that is almost 50%. Although, traditionally, people in Messolonghi were also working in agriculture and fishery, those sectors have declined over the last decades. Tourism is also not well developed in the area. Although Messolonghi seems left behind from an economic point of view, it is a city with rich history and culture. It has also many places of ecological importance, such as the Lagoon of Messolonghi, which is classified as an environmental park with uncountable winged and aquatic residents.

Coming back to the case study, the founders of the local hub knew each other from another active citizens' group they participated in the city and decided to transform the family house of one of the founding members to a local cultural hub. The organization acts at different levels, as it provides the community with a space to read, work and socialize in a daily basis, and in parallel it organizes a variety of public events and cultural workshops. The initiative has now three core members that work for its activities on a daily basis and a lot of volunteers that run a specific workshop or activity (Member of "Messolonghi by locals", Interview, February 2022). The people taking part in its actions are a lot, some on a daily or weekly basis and some as participants in the bigger events the organization implements, for example, the annual cultural festival they organize in the city. Most of the activities of the organization are self-funded, some are funded by the participants' donations, but the place is mostly preserved by "people's donations" (Member of "Messolonghi by locals", Interview, February 2022). They have received funding by European Commission programs for specific projects they implemented while having a mild economic activity, providing local tours and selling local crafts. One of the main outcomes of the organization's workshops and events is thematic maps of Messolonghi, which show specific routes in the area that locals and visitors can follow to approach a certain quality of the region, e.g. an environmental route that shows the most important environmental sites of the region, a route based on the history of the place, a route that unravels the artistic production in Messolonghi and others. The organization is also oriented towards the "organization of cultural tourism and ecotourism experiences for locals and visitors of the city," according to its website. As the members explain, they have a lacking collaboration with the municipality. They also recognize difficulties in the communication among the local actors in Messolonghi, and they try to actively bridge this gap, by organizing discussions between the municipality, its services and other local actors, on specific issues and problems the place faces. In those discussions, they try to initiate a participatory model through which the citizens can play a decisive role in the place's governance.

But what about the organization's members' motives? As stated in the organization's website, their main aim "is to strengthen collectivity within the local community of Messolonghi through a participatory mapping scheme of the city's cultural identity but also through the creation and promotion of cultural activities." And this is what they do the last 4 years, through citizen's mobilization, environmental activities, cultural projects, seminars, workshops and more. The organization was founded in a hard financial period for Greece, during which – as stated in the hub's website – "the social and economic crisis has increased the number of young people leaving and has created a feeling of introversion and inaction within the local community." As a member of the organization told us the idea behind the hub is simple:

> It started with a simple need, since we want to live in a place, we have to do something for it. There were vivid discussions every day in the coffee place. People would talk about the problems of the community, for example, the litter, the stray animals and so on. There was the discussion, but not the next step, which is to act. So, we took the responsibility to do something. (Member of "Messolonghi by locals", Interview, February 2022)

This feeling of inaction and apathy, that this member describes, are related to the ways rurality and peripherality are usually perceived and stereotyped, as places where "nothing happens". As it seems, the need to "do something", to overcome this feeling of inaction and incapability to act, is the first and main motive for the founding of this organization.

Creating Possibilities by Reimagining Development, Mobilizing the Community and Converting Identities

Reimagining Development

As the interviewees argue, when the people in Messolonghi are young, they are "brainwashed" to "leave this place" as "there is no future here" (Member of "Messolonghi by locals", Interview, February 2022). This attitude is a stereotype that is produced by but also (re)produces the marginalization of this region common characteristic in most perceptions of rurality and peripherality. Since the peripheries are – to some extent – "discursively produced" (Willet & Lang, 2018, p. 258), the attempt to overcome the "no future" stereotype for Messolonghi is a first step to create new narratives but also new possibilities for the development of the region. The interviewees first realize the problematic situation of the developmental pathways of their town. As they say, the town has been developed based on the universities that were placed in the area and the students' needs. The main financial activities in the region were accommodation renting and services for providing food and drink. But, as they argue, after the universities were closed, the place was found at a dead end:

> The economy was based on rooms to rent for the students and coffee shops. So we didn't have the vision to do something different. It was only about night life and where to live, where to stay. So, when the university left and the students left, we had coffee shops and airbnbs, nothing more. We try to change this. (Member of "Messolonghi by locals", Interview, February 2022)

This is why they are trying to promote alternative ways of development for their region. For example, they are shading light to traditional practices that have been neglected the previous years, such as fisheries and others. They are also trying to actively promote – also through the hub's gift shop – local agricultural products and crafts. Besides that, the members of the organization argue that they aim to illuminate Messolonghi as a place for ecotourism. They want people to consider Messolonghi as a place they can visit "with respect to the environment, the locals and the cultural heritage of the place" (Member of "Messolonghi by locals", Interview, April 2022). The initiative also attempts to show another approach to regional development. The organization, according to its members, aims to adopt a more holistic understanding on the ways a region is developed and as many scholars have pointed out (MacKinnon et al., 2021, Pike et al., 2007) pay attention also to political, environmental and social aspects of development:

> Even in our profit-driven activities, we try to follow the combination of economy, environment, society and not only make money. For example, in one action of ecotourism, you learn about the Lagoon, you respect it more, you become a more active citizen while taking part in an action, and you also support the fisherman, the professional, the tour guide and the organization. (Member of "Messolonghi by locals", Interview, February 2022)

"Messolonghi by locals" attempts to create new imaginaries for development by challenging the "no future" narrative, promoting different developmental practices and showing a holistic approach to development. The limitations of a small-scale organization to create a developmental strategy for a region are for sure many. Nevertheless, such an organization can – and attempts to – open the space for the discussion and the reimagination of development in the region.

Mobilizing the Community

The team also aims to mobilize the community. Through their cultural activities, they firstly attempt to enhance the feelings of attachment to the place and in that way strengthen the communal relations. As they say their main logic draws from "socioculture", which for them is the combination of cultural and societal elements and the introduction of culture into the everyday life of the community:

> We use a lot the cultural heritage and the place's history but what is important for us is how to unite the community. (...) Our heritage and our identity is about making someone walk on the street and feel good about the place they are in. Because in the periphery in particular if you ask the young people everybody wants to leave.
>
> Member of "Messolonghi by locals", Interview, February 2022

They are also trying to mobilize the community to make specific changes in the area. They attempt to raise awareness for the place's problems; they also organize collective activities to overcome those. As they say, while explaining an environmental action they planned and implemented in a road near the region's Lagoon, which is a site of great environmental importance:

> This road used to be filled with litter and also had a bad reputation, so people would not go there. We acted so that people would go there again; the place was cleared of litter, preserved and renewed, and became a route for birds' observation, a pleasurable walk. (Member of "Messolonghi by locals", Interview, February 2022)

As the interviewees argue, they are trying to activate people to do things all together for their place. As they say, "we are here and people know we do things for the place, so sometimes they pass and they mention problems, asking us to do something about them," but the members try not to cultivate this logic; they ask people "to come and discuss and act collectively on those issues" (Member of "Messolonghi by locals", Interview, April, 2022). As they say this was not easy in the start, but gradually it has become more rooted in the local population. Also, they are trying to activate other collectives and local actors in the region to change the way people relate to each other and to the place surrounding them. As they say:

> Also other collectives and the municipality took action after the issue [the environmental problem in the Lagoon] became public. It is not that we go alone to do it, we go all together to do it. (…) This was a place that very few people went to, and now people are walking, having picnics, children playing, etc. So, everybody wants to help in its preservation. (Member of "Messolonghi by locals", Interview, February 2022)

"Messolonghi by locals" attempts to create new possibilities by strengthening the communal bonds of the locals and mobilizing the local community and other actors to collectively act to change the problematic situations of their place. There are many questions that arise from these observations, such as the importance of the physical space of the hub, the relation of the physical space and the sense of belonging of the people and the importance of people acting and not just assign tasks to collectives and organizations of "experts". Another important question is how such organizations can find a fine line between mobilizing their community and replacing local actors by taking initiatives they should take.

Converting Identities

The mobilization of the community, as part of the everyday life (Gibson & Graham, 2004, p. 29), has also effects on a personal level, in the identities of the people involved. As the interviewees put it, they try to achieve small things at a time:

> If you have a strategy this could also lead you to small victories until your target is reached. (…) You see that when you persist and do something you can also have results; you achieve something (…) And people feel part of the change. (Member of "Messolonghi by locals", Interview, February 2022)

As the question of identity is so much bounded to the question of space (Massey, 1999), to create new possibilities for a place, you have to create new possibilities for the locals' personal identities as well. You have to dis-identify from the hegemonic subject position and cultivate a counter-identity (Gibson-Graham, 2006, p. 77). This is also another objective for "Messolonghi by locals". And they create a lot of new opportunities for people to transform their identities by "expanding their horizons through culture and arts," meeting people they would not have met otherwise and even "getting in touch from people from other ages, cities or countries" (Member of "Messolonghi by locals", Interview, April 2022). All these encounters are according to the members of the initiative really enlightening for everyone involved.

By initiating a more positive approach to life in Messolonghi, the organization creates the conditions for identities of people to change based on positive feelings.

As Gibson-Graham argue, through psychoanalysis we have recognized that the attachment to ones' identity is "cemented by enjoyment and frustrated desire" (Gibson-Graham, 2006, pp. 129–130). So, it is – also – this pleasurable relation that needs to be cultivated in the creation of alternative and counter-identities. In the activities of "Messolonghi by locals", we can identify this pleasurable relation between the people and a joy that comes from collective action (Ruddick, 2010). One of the members I interviewed gives us more information on this part. As she says "this is creative for me", and she goes on saying about the meeting they would have in the hub the night after our interview, "we will meet, laugh and prepare the games for our next actions, this is also fun, it comes from inside, this good feeling when we do things together for our town" (Member of "Messolonghi by locals", Interview, April 2022). Or as another interviewee argues:

> It is a ray of light. If you have the chance to read a local newspaper [he shows us the newspaper]: troubles, troubles, troubles, problems, problems, hate speech (…) and then it comes a reference to our action which is actually a ray of light. You know [he reads the newspaper's headline] 'Sundays become interesting in Messolonghi', so it is a completely different approach to life. (Member of "Messolonghi by locals", February Interview, 2022)

"To change ourselves is to change our worlds" as Gibson-Graham aptly argues (Gibson-Graham, 2006, p. 127), and this is part of what "Messolonghi by locals" attempts to achieve by its activities. As shown above, the organization aims to create new possibilities through politically empowering the locals, offering them broader encounters and creating a pleasurable identity for them in relation to the place.

Conclusions

In this chapter I approached a local hub as a space that can potentially create new possibilities for the region of Messolonghi and beyond. I focused on the objectives of the members of "Messolonghi by locals" and the ways they attempt to create social change and new possibilities for their community. I tried to better understand the people's involved points of view, their strategy, their aims, the logic behind their activities and so on. Through this analysis I conclude to a categorization of the different ways such a space of meeting and collaboration can attempt to challenge the existing situation and create new possibilities.

The objective of the people founding and running "Messolonghi by locals" is to create social change in their town and region in a variety of ways. Those range from a wider and more strategic scale to a more personal one. I argue that the categorization and deeper analysis of those ways can give us clearer insights into how new possibilities can be created for places. "Messolonghi by locals" attempts to create new possibilities by reimaging the development of the city and the region and more specifically by challenging the stereotype of the underdeveloped periphery, by promoting new developmental practices and by approaching development in a more holistic and progressive way. It also attempts to create new possibilities by mobilizing the local community, which means firstly to strengthen the feeling of attachment of people to the place and the communal feelings among them, secondly to transform the place itself in a physical way and lastly to mobilize the community to act collectively and address the place's problems and challenges. Last, the organization opens up Messolonghi and its community to new possibilities by converting the identities of the participants and more specifically by empowering the people involved, by exposing them to new encounters and by creating new pleasurable identities for them in relation to their place.

There are a lot of new questions that these observations open for further research. Firstly, it is worth to deeper research the potentials as well as the limitations of such a small-scale initiative in contributing to the developmental strategy of a town or region. In which ways could they most effectively collaborate with other local actors? There are also a lot of questions arising regarding the relation of physical space and the community. How important is that the hub offers a physical space for everyday encounters? What is the relation of spatial transformations to the feeling of belonging of a specific community? How can such an initiative avoid cultivating the logic of assignment to people? Since the organization obviously offers things to the community that the municipality should offer, how can they avoid replacing local administration and taking care of their responsibilities? Lastly, could we identify similar structures and lines of thought and action between other spaces of collaboration in Greece and beyond? Can they help us to imagine new ways to mobilize communities and creating new possibilities for a place or region?

As stated in this chapter, "Messolonghi by locals" aims to create new possibilities through rethinking on a regional level, the development of Messolonghi; on a local level, the place and its community; and on a more personal level, the identities of people involved. Through this approach the strategy of the organization is not only looking at the big picture, nor only focusing on the everyday life of the place. There seems to be a holistic understanding that leads to action at different levels and scales while opening up new possibilities for collective and personal futures. This local hub's activities apart from giving us insights for ways to create new possibilities give us also insights on going from the strategic to the personal and vice versa. It can only remain an open question for the purposes of this chapter how those different levels interconnect and if this model can be helpful for the initiation of a bottom-up approach of how we can live and work together in a region.

Acknowledgements This paper has received funding from the European Union's Horizon 2020 research and innovation programme under the Marie Skłodowska-Curie grant agreement No 955907, project CORAL (Exploring the impacts of collaborative workspaces in rural and peripheral areas in the EU).

References

Avdikos, V., & Iliopoulou, E. (2019). Community-led coworking spaces: From co-location to collaboration and collectivization. In R. Gill, A. C. Pratt, & T. E. Virani (Eds.), *Creative hubs in question* (pp. 111–129). Palgrave Macmillan.

Avdikos, V., & Kalogeresis, A. (2017). Socio-economic profile and working conditions of freelancers in co-working spaces and work collectives: Evidence from the design sector in Greece. *Area, 49*(1), 35–42.

Avdikos, V., & Merkel, J. (2019). Supporting open, shared and collaborative workspaces and hubs: Recent transformations and policy implications. *Urban Research & Practice, 13*(3), 348–357. https://doi.org/10.1080/17535069.2019.1674501

Avdikos, V., & Pettas, D. (2021). The new topologies of collaborative workspace assemblages between the market and the commons. *Geoforum, 121*, 44–52. https://doi.org/10.1016/j.geoforum.2021.02.012

Brinks, V. (2022). Fluid objects? An attempt to conceptualise the global rise of "coworking spaces". *Area, 00*, 1–8. https://doi.org/10.1111/area.12778

Capdevila, I. (2013). *Knowledge dynamics in localized communities: Coworking spaces as microclusters.* https://doi.org/10.2139/ssrn.2414121.

Capdevila, I. (2015). Co-working spaces and the localised dynamics of innovation in Barcelona. *International Journal of Innovation Management, 19*, 1–25. https://doi.org/10.1142/S1363919615400046

De Peuter, G., Cohen, N. S., & Saraco, F. (2017). The ambivalence of coworking: On the politics of an emerging work practice. *European Journal of Cultural Studies, 20*(6), 687–706. https://doi.org/10.1177/1367549417732997

Dinardi, C. (2019). Grassroots creative hubs: Urban regeneration, recovered industrial factories and cultural production in Buenos Aires and Rio de Janeiro. In R. Gill, A. C. Pratt, & T. E. Virani (Eds.), *Creative hubs in question* (pp. 299–317). Palgrave Macmillan.

Gandini, A. (2015). The rise of coworking spaces: A literature review. *Ephemera, 15*(1), 193–205.

Gandini, A., & Cossu, A. (2021). The third wave of coworking: 'Neo-corporate' model versus 'resilient' practice. *European Journal of Cultural Studies, 24*(2), 430–447. https://doi.org/10.1177/1367549419886060

Gibson, K., & Graham, J. (2004). The violence of development: Two political imaginaries. *Development, 47*(1), 27–34. https://doi.org/10.1057/palgrave.development.1100013

Gibson-Graham, J. K. (2006). *A postcapitalist politics*. University of Minnesota Press.

Lee, D. (2019). Creative hubs, cultural work and affective economies: Exploring 'unspeakable' experiences for young cultural workers. In R. Gill, A. C. Pratt, & T. E. Virani (Eds.), *Creative hubs in question* (pp. 69–88). Palgrave Macmillan.

MacKinnon, D., Kempton, L., O'Brien, P., Ormerod, E., Pike, A., & Tomaney, J. (2021). Reframing urban and regional 'development' for 'left behind' places. *Cambridge Journal of Regions, Economy and Society, 2021*, rsab034. https://doi.org/10.1093/cjres/rsab034

Mariotti, I., & Akhavan, M. (2020). Exploring proximities in coworking spaces: The evidence from Italy. *European Spatial Research and Policy, 20*(1), 37–52. https://czasopisma.uni.lodz.pl/esrap/issue/view/427

Massey, D. (1994). *Space, place and gender*. University of Minnesota Press.

Massey, D. (2005). For Space. London: SAGE Publications.

Massey, D. (1999). Philosophy and politics of spatiality: Some considerations. The Hettner-lecture in human geography. *Geographische Zeitschrift, 87*(1), 1–12. http://www.jstor.org/stable/27818829

Merkel, J. (2015). Coworking in the city. *Ephemera, 15*(2), 121–139.

Merkel, J. (2019). 'Freelance isn't free'. Co-working as a critical urban practice to cope with informality in creative labour markets. *Urban Studies, 56*(3), 526–547. https://doi.org/10.1177/0042098018782374

Pike, A., Rodríguez-Pose, A., & Tomaney, J. (2007). What kind of local and regional development and for whom? *Regional Studies, 41*(9), 1253–1269. https://doi.org/10.1080/00343400701543355

Pratt, A. C. (2021). Creative hubs: A critical evaluation. *City, Culture and Society, 24*, 100384. https://doi.org/10.1016/j.ccs.2021.100384

Ruddick, S. (2010). The politics of affect: Spinoza in the work of Negri and Deleuze. *Theory, Culture & Society, 27*(4), 21–45. https://doi.org/10.1177/0263276410372235

Schmidt, S., Brinks, V., & Brinkhoff, S. (2014). Innovation and creativity labs in Berlin – Organizing temporary spatial configurations for innovations. *Zeitschrift für Wirtschaftsgeographie, 58*, 232–247. https://doi.org/10.1515/zfw.2014.0016

Spinuzzi, C. (2012). Working alone together: Coworking as emergent collaborative activity. *Journal of Business and Technical Communication, 26*(4), 399–441. https://doi.org/10.1177/1050651912444070

Vidaillet, B., & Bousalham, Y. (2020). Coworking spaces as places where economic diversity can be articulated: Towards a theory of syntopia. *Organization, 27*(1), 60–87. https://doi.org/10.1177/1350508418794003

Waters-Lynch, J., & Duff, C. (2019). The affective commons of coworking. *Human Relations, 74*(3), 383–404. https://doi.org/10.1177/0018726719894633

Willett, J., & Lang, T. (2018). Peripheralisation: A politics of place, affect, perception and representation. *Sociologia Ruralis, 58*, 258–275. https://doi.org/10.1111/soru.12161

Part IV
Epilogue

A Conversation with Ashley Proctor and Bernie Mitchell

Janet Merkel, Dimitris Pettas, and Vasilis Avdikos

On the 8th of March 2023, the editors had an online interview about the current state of the coworking movement with two notable figures of it, namely, Ashley Proctor and Bernie Mitchell. We discussed various issues and challenges that coworking spaces face nowadays, and the reader can find direct links with some of the arguments that the previous chapters address. We truly thank both for giving us their valuable insights on the evolution of coworking and its potentials for workers and communities.

Ashley is one of the original voices of the coworking movement, developing collaborative communities since 2003. She has launched the Xpace (2004), Creative Blueprint (2006, 2012, 2015), Foundery (2011) and 312 Main (2018) coworking communities, artist studios and gathering spaces in Toronto, Vancouver and Seattle. Ashley is one of the founding members of Coworking Canada, Coworking Toronto and Coworking Ontario and has been leading these alliances since 2010. In 2013 Ashley created COHIP—the world's first Coworking Health Insurance Plan. COHIP is now available to all independent workers across Canada. Ashley co-produces the Coworking Alliance Summit and manages the Coworking IDEA Project.

Bernie has been active in the coworking industry since 2010. He passionately advocates freelancing, remote work and 15-minute city movements. Bernie is co-founder of the London Coworking Assembly, where he works on the Coworking I.D.E.A. Project and Urban MBA. He is also a founding partner of Velvet Community, which works on fintech products within the coworking and co-living industry.

J. Merkel (✉) · D. Pettas
Institute of Urban and Regional Planning, Technische Universität Berlin, Berlin, Germany
e-mail: janet.merkel@tu-berlin.de; d.pettas@tu-berlin.de

V. Avdikos
Department of Economic and Regional Development, Panteion University, Athens, Greece
e-mail: v.avdikos@panteion.gr

J. Merkel et al. (eds.), *Coworking Spaces*, https://doi.org/10.1007/978-3-031-42268-3_15

E: *Could you please just briefly introduce yourself and what you're doing in the last few years within the coworking movement?*

A: I'm involved in so many different organizations in coworking. It's been my entire career for the past 20 years now. Only coworking, the only thing I've ever done. My main company is Creative Blueprint and what we do is consulting and project management, and the projects that we work on are usually collaborative or cooperative projects with other coworking operators or ones that are designed to support the coworking movement. Like COHIP, for example, the Coworking Health Insurance Plan, which I've been managing for the last 10 years for all independent workers across Canada. This was born out of a coworking community. I run Coworking Canada, which is an alliance of independent operators across Canada and collaborative organizations that want to help make work and the world a better place. The Coworking IDEA Project, which is an international collaborative organization that Bernie is also a part of, designed to ensure the coworking movement and industry remains and becomes more inclusive, diverse, equitable and accessible. I also produce the Coworking Alliance Summit which is designed to bring the world's coworking alliances together to share their practices and to discuss some of the topics and interests we share more deeply. So, these are the organizations. And I'm an activist and an artist.

B: The story I'd like to tell you is I got into coworking in 2010, when I got hired by a coworking space incubator in London to be their event guy. And before that I was running networking groups. And what I found in coworking was what I was looking for in the networking groups. Because when we had the networking groups, people would come along and go, you know. Hi, I'm Brad and I sell this, and everyone was in like engagement mode. And then when we got into the coworking space, everyone chilled out and started talking to each other and helping each other out. So, I've always been getting people together. And then in 2014, I went to coworking Europe for the first time, and I'm sure I met Ashley there. When we came back from that to London, we started what is now the London Coworking Assembly. And pretty much nonstop since 2014, I've been getting independent, not only independent coworking spaces, but mainly independently owned and run coworking spaces together to help each other out, to learn from each other. I've consulted and worked with coworking spaces on their marketing and communication and stuff like that. Over the years I've learned more and more about it, and now I learned how to work with people on their business and how to keep their coworking space alive. Supposed to integrate up, we call that coworking consulting. But I'm really interested in getting people together and business as a force for good rather than, you know, sucking the blood out of everything.

E: *How is coworking evolving in the last few years? What has changed? What about some new trends, like the increase of digital nomads and the co-living spaces? How does the industry evolve and especially for those more independent coworking spaces?*

A: I've witnessed the evolution of the coworking movement. So, starting with those, you know, independent, community-led and focused spaces around the world, building up to what it is now, and the definition expanding as well. And that's something we also predicted, the word coworking actually changing over time, and the way we use it. I think we're seeing now what we call coworking will just become work. And everyone will do it that way. That's the goal, at least. But I think we've seen it evolve from community-led, small spaces that are really focused on the values of the movement and dismantling loneliness and the collaboration over competition. We've seen it evolve over time into an industry, and that was perpetuated with the rise of brands like WeWork really commercializing the intention behind that and turning it more into a real estate play. The industry really boomed after that though, with all the other supportive technologies to make coworking more possible for more people. And then of course, info and tech workers now obviously have more opportunity to cowork from everywhere as digital nomads. But I think we're still in that gap between the privileged few having access and moving into a more mainstream act, and really, I see coworking as an act, it's a verb, it's a thing we do. And I think it's becoming more normal. We're starting to see the values of the coworking industry, and movement spread out amongst other industries and start to influence and impact other industries as well who are now working more collaboratively. I think the pandemic accelerated that shift, but I think coworking has always been there. We are the essential infrastructure now for this new work landscape. And what we're seeing today, the solutions that we're coming up with today for remote and distributed work wouldn't be possible if coworking communities haven't laid that essential foundation prior to the pandemic. I've seen a lot of changes. I mean, there are so many other changes and evolutions, obviously more corporate teams and larger teams. It's not just for the entrepreneurs and freelancers or small businesses anymore; that's the past, you know; I'm signing up teams with hundreds of thousands of employees, and thousands of employees spread out around the world and they're just giving everyone access to coworking. So that was not happening even a couple of years ago. So, there are lots of shifts across the industry.

B: I think it was 2016, my friend Neal did a book launch. I was the coworking guy on the panel, and I asked the audience, what is coworking? And because I didn't want to lecture and to be honest back in 2016, I wasn't sure what coworking was. So, I wanted to ask the audience, and people went around and a lot of people would be into coworking in Europe, and you know, knew that different people were going on. And there was the Center for Social Innovation in Canada. Canada was one of the places people mentioned. And then someone at the back said, oh, it's what big corporations are re-washing their services at. They said it's what people in corporations are, you know, dressing up their offices at. And then Juliette who worked for British Land at the time said, "you know, I'm one of the wankers from the corporate sector that's doing this". But then she went into this is what people are coming to ask us for. And she said, "you know, we're calling it that because this makes more sense from a sort of designing spaces that are not open plan but collaborative and people can move". And now people come and ask us for. They say, can we have a coworking space, and I was a little bit devastated to hear this because I thought that the evil corporate sector was just doing this. So, companies and people wanted not a new way of working, but a different way of working or a more connected way of working. So that was one thing. Also in 2016, I first heard the word co-living, and because I've seen coworking come around, I was like, oh my goodness, I'm gonna pay attention to this because then I can be on the co-living wave. And I went to be part of a group called I think Weshare which is something that came out of the sharing economy in Paris. And I went there, and I thought it would be full of annoying people in white clothes and tattoos, talking about being a digital nomad. And what I found, and it was significant, was a group of people that were very concerned about how we're going to live in cities and work together. And there was this guy, he said if we live separately, like all the old people over here and all the young people over here, we're going to miss out on learning from each other. And when people live together, they live longer. So, when older people and young people live together, everyone lives longer. And one of the other threads that came out of there, which touched on it is that, you know, the whole thing about privilege and what I think has happened is companies like WeWork and such for they have massive budgets, and they appeal. When you go to Google, Second Home, the Office Group, who are like WeWork, and a couple of other places in London, they dominate the voice in the marketplace, and you don't hear about the Alice's corner coworking space and stuff like that. So, everyone's perception is, "WeWork did a lot for coworking", and it's like saying MacDonalds did a lot for restaurants. You know, they have massive budgets, they bleed into the marketplace, and they spread the activity, but they don't really miss out on the essence that's available through coworking. But the other thing I need to say is I've read a lot about WeWork, and I've never met a nasty person that works in there and I was just running a panel in London with the head of design for WeWork in Europe, and I was apprehensive about it, but they were just like good people who work for a corporation. So, there's something there, but I don't like that kind of scorched earth mentality that comes with running a huge corporation. I think that one of the things I'd like to see, and I think it's happening more because of COVID, is people staying in their neighbourhoods and that 15-minute city – more coworking space becoming part of the local area and creating more connections in the local area, that kind of thing.

A: Yeah, I think we should also note the trend towards specialization and targeting niche markets, specific membership or target market. Also, the increase of purpose-driven community-led spaces. So, it's not just a target market, but it's a sort of an activist or advocacy lens. When we're elevating certain groups of people, coworking spaces for women, for example, I think that the work that we've been doing over the past two decades and all those members that have come through coworking spaces have been influenced and impacted over that time as well. And they're leading companies, and they're going out in the world, and working at companies and making decisions, and they're running households, and those folks are acting more collaboratively having been part of a collaborative space and network. And so, we're not just seeing coworking spaces shifting, we're seeing the way that cities operate, we're seeing the way that other corporations are operating. We're seeing more collectives and alliances and supportive structures in place. I think thanks to coworking as well. So, it's not just the spaces that are changing, but I think we're seeing a ripple effect in everything that coworking touches and all the people that coworking impacts.

E: *What do you think are the challenges for the more independent ones? How do they act in the market? How do they react to all these changes?*

B: In London, which is where I know those spaces the best, there is a confidence thing because there's like space A saving space B, and they're pretty much the same. One space will be terrified by WeWork and the other space will be like, I'm just doing my thing. Someone posted that there are 1300 coworking spaces in London, and it depends on what people consider as a coworking space. But the reality of the situation is you would think there's only those five big brands available to cowork in London, but there's a coworking space near King's Cross for ex-offenders to rehabilitate and get to the world of work. There are so many people who come to London Coworking Assembly who either run a space on their own or run a space with a business partner. They are photographers and they run a coworking space, they are a design agency and they run a coworking space or there's a space which is part of a co-op. And there's the Affordable Workplace scheme which is particularly strong in Islington. There's all these other spaces and it's definitely a reality. They're like competing with the big ones. We know that it's always a bit of a comparison of coworking spaces and the restaurant industry. So, you know, you're competing against this mega Gordon Ramsay and McDonald's versus Ashley's coffee shop. And there is always that tension which I suppose in some ways it's just the reality of business, but sometimes it plays out unfairly in people's lives.

A: I think that one of the challenges is an advantage as well. So as an independent operator, you might not have the same scale, and scalability is important in business. So, you know, if we're opening locations in 50 cities, we're going to have some potential that a small operator doesn't. We're also going to be able to maximize our profitability at that kind of a scale and a small location or a single location isn't necessarily extremely profitable, but that's usually not the intention for these folks running these spaces. And like Bernie said, they may have other jobs. And that's definitely a challenge, but really it is, you know, as a woman or as an independent operator, starting out a new coworking space. I literally won an award for my coworking company. And then the bank that presented the award turned me down for a very small business loan because they said this will never take off. "People have worked their whole lives for the corner office. Why on earth would they give it up to work in a shared space with other people?" This was the same bank that gave me the award on stage, but behind the scenes refused to invest in an idea like this, which they didn't see as scalable or successful. So, access to capital is a challenge. And the size minimum for profitability is also a challenge. You know there's a minimum size that makes sense to operate a space like that. Otherwise, you're just sort of operating a non-profit. And that might be the intention again, you know, to collaboratively make this space accessible to more people. And many folks do that. I think it's hard, but I think there are real advantages as well as being those small spaces, and Bernie touched on it with the collaborative elements. And so, of course, independently, I'm not going to be able to take on WeWork because of the massive budget. If I am trying to access some of that technology, they may have an advantage and be able to develop technology, instant booking, and things like this across platforms to make it easier for their clients to use it. But as an independent, we also work together. We can participate in things like ONE Coworking or Deskpass and collectively collaborate, and we collectively as independent operators make up the majority of all the coworking spaces in the world. And when we work together through platforms like Deskpass or ONE Coworking, for example, we are more appealing to someone looking for a network of coworking spaces because you're not going to go to a Starbucks in every city. You want to go to "Ashley's little cafe" in each city and have that unique experience. And so, we have a superior product in terms of the differences, in a real experience in a new city, in the same way you would experience McDonald's or Starbucks in every city or hotel chain, having the same experience in every city. That's not what we're trying to achieve. And for most people that's not what they want to experience in their workplace.

B: There's a capacity thing as well because, you know, like nearly 10 years I've been going to conferences. And there's huge value in showing up to any of these conferences. But if you run that business, you can't like the mental risk of going. I'm gonna leave my coworking space in the hands of, you know, Jackal Jenny, the community manager and go to Lisbon for four days and hang out with John and Ashley and Bernie and the gang. And it's like, why am I going? At the Workspace Design Show Conference, there were bigger companies, like four people from each company there. Whereas for the independent coworking people came in, did the London Coworking Assembly event and went and we got a lot of people to come to the workspace design. So because it was in London, it was free to go to; whereas other conferences because they have a different business model, it's like £500 to go to Coworking Europe plus your airfare, plus the time out of business and all that type of stuff so that there's a capacity thing there. And keeping up with probably thinking you need to keep up with technology when maybe you don't need to keep up as much as you think. I think people sometimes read an article on a medium that says coworking spaces taking Cryptocurrency now and then they panic that they're missing out and then they don't have time to research that actually it's one block-chain thing in Sweden that's managed to do it run by someone like Yang, whereas everyone else is just still taking PayPal payments, you know, like the panic of missing out is because I haven't got the resources is something else too.

A: Yeah, I think that all goes back to the essence for the alliance, our collective for the smaller operators because the larger teams and you know, multi-location, they may have 40 community managers and you just have a staff meeting and they're all solving problems and coming up with efficiencies and sharing things they've learned or feedback from other clients in different markets. And, and if you're one operator in one small city, you don't get that at all. But we've created these networks, and as an operator in Toronto, I can participate in Coworking Toronto Collective, Coworking Ontario Collective, Coworking Canada. And I can attend all these international conferences where we're all sharing our ideas and practices, and that's not industry standard anywhere else, but it is in coworking. And so, I think again, that's an advantage of being one of the independents, although there are still a lot of issues and struggles. I think we've done an incredible job of showing folks that you don't need to be the biggest player to be effective, that you really can just be small and impactful in your own community. And as you know, all the little guys that make a difference.

B: One last thing on that is I think there's nothing else when you run a business on your own. And I think particularly with a coworking space, you're like the person that runs the coworking space and you have maybe a couple of staff and then you have all these people, and you accidentally end up being like someone who's renting a desk, someone who's a business coach, someone who's a community manager. It's almost like being a minister of religion. You have this whole little tribe to take care of our congregation. And I think that is also mentally exhausting and there's a lot going on, and I think there's a loneliness and mental health wish that comes with that until you get smacked in the face by it, you're not available, you're not aware of it until it brutally reveals itself.

A: I think there's a real risk of operator burnout and community managers, because of the emotional labour that's required to manage a coworking community. They're providing care constantly, it's a caregiving role in addition to a management role, and we don't compensate people the way they deserve for that kind of work. It's essential work. And we saw that over the pandemic, you know, entire communities relied on these folks to hold them up for years.

E: *True, we find a lot of invisible labour in coworking spaces. There are also several academic papers about labour burnout of coworking managers and community managers and the effort that you need to put in maintaining a coworking space and the coworking ethos and spirit. But we would like to ask you about some other kinds of challenges that smaller coworking spaces face, like the rising real estate prices in large cities and metros. How do coworking spaces cope with these?*

A: I think there's a few creative solutions. I think just by being a coworking space, you're enabling a lot of other businesses that couldn't compete in that market to compete by sharing the cost. So just in its essence, it's an opportunity. But as an operator there are a lot of opportunities to partner. I think you can partner in terms of a management agreement with a landlord. You can partner with a developer, but you can also partner with the city or a non-profit organization and do it for a social purpose as well. A lot of the coworking spaces that I've built and managed over the years are directly funded by the government. The city provides an unused asset, and you know, if we're doing the renovations and improving economic development in the area and making sure that there are returns for the city and the neighbourhood, then they're providing the space at a dollar a year lease, for 50 years to ensure that we continue to do that good work in the community because that's so attractive. They don't have to pay for it. And we come in and do the work for them. There are a lot of ways to get around the "per square foot" leasing model.

E: *This is the case in Toronto, but in other areas, in other cities, this is not the case because the state does not pay any particular attention to coworking. So, we probably have geographical variations and different policy responses to coworking.*

A: I fully agree there are regional geographical differences there. But I also want to say that for 18 years, that was true in Canada as well. And we pushed for it, and we lobbied for it, and we changed the definition of "a space that deserves to be funded". Not just creative spaces, not just cultural spaces, but also coworking spaces; membership based, collaborative spaces nowadays are eligible for tax discounts. And, that didn't exist, in the same way health insurance for entrepreneurs didn't exist. It was not allowed in that form, there wasn't even a way to do it at the time, but we pushed for 10 years to get it. So, I want to acknowledge the differences regionally, but I don't want people to say, "oh, that's true there, but it could never happen here". It couldn't happen in Canada 2 years ago. But now it is. And there's 300 million set aside for coworking spaces to ensure that they're going to exist in communities across Canada. And so, I want people to also know that these big things started very small with one person or an idea or a collective of people who care and keep pushing it, advocating for it over time.

B: I might have positive and negative things to follow with that. So, I use London again because that's where I know the most, and as a result of the pandemic the local authorities in London paid more attention to coworking than ever before, because as you know there are zones one to six. And I haven't got the data for this, but like 70% of the coworking spaces in London were in zone one. And then probably, another 10% in zone two and then scattered around... And where I lived on the east of London, we lived in zone four and for like 10 years I was looking for the coworking space, nearest zone four in, it's Elford, which I think you might know. And then suddenly the pandemic happened, and more alert local London authorities were like if we make people stay in our area, they'll spend their money in our area, our community goes stronger, and all that kind of stuff. So, there was a response by a lot of local authorities, and particularly around southwest London, it was like there's a thing called the Big South. They were like right out the gate. And it's a collaboration between universities, developers and local authorities, and it's led by local authorities. They're very alert to doing similar things to what Ashley described. But then on the negative side, people are just very scared that they've got a coworking business with tight margins and the cost of energy goes up and like how to message it: "we need people to come in to keep making money, but then they're gonna use more electricity, so our costs will go up and our revenue will go up". And it's just like a nightmare balancing act. And you would've seen on LinkedIn. As a result of the cost-of-living crisis, should we tell people that it's better value for them to pay us to come in and use our electricity than stay at home and use their own electricity? So, I think I haven't got the answer for it, but this is the worst situation in my lifetime I've ever seen economically in the UK, and people who were trying to work it out anyway before the pandemic are now like punched in the face by the state of the government and the state of the economy. And one of the good things to come out of this is how big companies are just ripping off the nation as a result of the way the economy is structured. So that's what I mean, that's not a positive thing, but that's one of the things that people were realizing, people who didn't really need to worry before, like nice white middle class people, like, oh my God, we need to get together and stand up for something. Whereas we haven't really had to stand up for anything in the last 30 years… that kind of mentality. So, there's a kind of forced collaboration for first welcome trees becoming available.

A: I think the pandemic experience has forced everyone into this realisation that a lot of the systems that were in place previously are broken or weren't designed for us in the first place. And just the opportunity to reflect, to slow down, to really understand some of the complexities and to see how we function in a crisis, I think it really threw a lot of people who thought maybe we would all come together, and some groups really did, and they succeeded and they're thriving today because of that. And it's what got folks through. But some folks really went towards independence, and we see this individualistic streak, particularly in North America. And it really sort of was a perfect use case to prove that that wasn't the successful way out of this. The only way to go forward is together. And the only successful models we saw come out of the pandemic were collaborative ones. And I think the only things that we'll see thriving in the future, you know, post-recession, post pandemic, are the ones that are collaborative as well. And as a movement, coworking has been through this before. The big push, like when WeWork and everything else was born, came out of the last recession that we went through in 2008. And people were freaking out at the time; "this is a terrible time to launch a business. What will I do with my coworking community?" And yet, small businesses were formed at an astonishing rate. Entrepreneurship and self-employment boomed, and those few coworking spaces that existed were packed full of new entrepreneurs during a time where most buildings were empty, leases weren't being met. The coworking spaces were the most secure bet during that time. And it really flipped people's idea of precarity. You know, we used to see a job as being secure and self-employment as being precarious. And yet in that last recession, we saw a whole bunch of people go, "wait a second, that job is very precarious and the only thing that's actually secure is my self-employment and my self-directed work. And when I do that in community, I'm really, really minimizing my risk and maximizing the opportunity I have to succeed in this kind of a climate". And we're seeing that again today. So, it's very familiar. It feels very familiar.

B: Tom Peters has been saying for years that we need to be the CEO of ourselves. And his vision came true. There's a place called Good Space, which is in London. And David launched Good Space in August 2020. And part of me was cheering him on! But another part of me was like, oh my God, what's this guy doing? And he opened a neighbourhood coworking space, he called it a neighbourhood workspace, and his heart is in the coworking side of it. And in like a month he got 65 members. It just filled up with people from the local area and he didn't do a vicious advertising campaign. He just had a board and stood outside his coworking space and was lucky because people had to walk past it. And I think he brought something to that little corner of London that people didn't even know was available. He brought this neighbourhood workspace; people came together because of the way these big double doors opened. So, he had them open and people could easily put their head inside and his whole business took off. Other examples are available. There is the Urban MBA with which we do a lot of work in the London Cowork Assembly, which is a youth organization. People go through this, young people learn how to start a business, start their freelance practice. And lots of people who were part of Urban MBA over the last five years turned up to that community and were like, you know, can you help us out? And one of the shining examples there is a, a single mom called Rachel who had a little bit of a food business going at the weekends, just became a full-time mobile chef. And she started getting online cooking TV shows. So, she's gone from a corporate job, doing a little bit of cooking for parties at the weekend, to full-time independent chef TV star. And there's loads of stories like that, which is really encouraging, and there are things we did in the London Coworking Assembly to connect people to make that happen. Which is really cool.

E: *Do you think that coworking offers a kind of emancipatory potential through the coworking practice for freelancers and for owners of small businesses? And to what extent? Is it about solidarity and care practices? Is it about organizing? Is it about organizing upon political claims or about the future of their neighbourhoods?*

A: Well, I think that we talk about work a lot when we're talking about the potential of coworking, but I don't think of coworking as strictly related to the ways in which we work. And so, I see it as sort of a way in which we live and operate and approach everything. I think coworking has the potential to transform absolutely everything we do. I want to be clear that this isn't a new mindset or practice. Obviously people have been coworking forever, and I want to acknowledge Indigenous people, coworking women particularly over the years, coworking to complete chores, coworking and hunting even, in the past there are so many different examples of people coworking and collaborating over time. But this is a really intentional act. And I wanna just clarify that I think that that act of coworking intentionally is what transforms things. And I think it's not only creating an opportunity for people to be supported as independent workers, but I wanna stress the interdependence as well. And I think that's a big key. We wouldn't be able to provide the Coworking Health Insurance Plan for all independent workers in Canada, if we all weren't working together on it – if we set up five different plans, none of them would be able to offer what we can offer when we work together to get rates with our numbers across Canada. And so, it really takes that collaborative element so we can scale, and we can grow in a positive way as well. But I think it empowers people not only to be able to work remotely. But I think there are benefits for companies and there are benefits for the city as well as for the individual and the coworking community. And I think that in addition to enabling the distributed workforce and empowering them, it really helps with talent retention for the company. It helps with talent retention for cities as well, instead of people moving out of the city, to be able to live and work in the same place.... your home. Keeping local talent. And that really allows people to choose where they work, when they work, how they work. It allows people with disabilities to participate in the workforce working remotely and in community in this way, it allows working parents to contribute in a more meaningful way. All of this autonomy is really essential. And I think coworking again, is that infrastructure, is that network that really supports all of this.

And so, in turn, we're building these resilient cities and all the things we're doing in the community by presenting the opportunities to connect and collaborate, job creation, increasing that local spend and all the economic development initiatives that the cities are trying to achieve... But we are succeeding in our spaces and that really allows people to show up in a different way. They show up more fully in a workspace that's designed to be inclusive and accessible. And you share a very small portion of the ideas you have in an environment that's not welcoming and inclusive and accessible. But if we're creating these intentional environments and we're collaborating, we're encouraging people to show up, share their best and do more. And I think measuring the impact on the members over time, we're seeing people growing at a rate they've never been able to achieve before. And success is defined in different ways. But I think that these folks consider themselves to be extremely successful in living balanced lives and being autonomous or self-employed. Even the folks that are working remotely, they're adopting the coworking culture and the strategies that the independent workers have. And that's actually influencing corporate culture today. And I think that's one of the things that we can take credit for. Not only are we empowering and supporting those independent workers and making it safe for them to be self-employed if they can still care for their families and then they can still have health insurance, for example. But it also allows them to take things up to a level that they couldn't possibly do on their own, by being in community. And it provides another level of connection and security.

B: I've experienced that first-hand and I believe it can. One is Fernando from Lisbon who has been doing this PhD for like 300 years now, and it is about the micro-learning that happens in coworking spaces. And when I met him first and he was doing that, I then went back into space, and I was sitting next to a German guy called Nils, who was a hardcore WordPress developer. And I would be googling how to put a picture in my WordPress site. And he would look over in this impatient way going, what are you trying to do? Um, press that button there. And he would save me hours just by looking at my screen and going "press that". And then I became alert of how much not going and looking for free help, but like, how much just comes to you? Like in the kitchen in Maynard there was Eleanor, every morning she used to say, oh, good morning, what are you working on today? And so many times I'd have no idea what I was working on today. I didn't even know where I was going. And because she would ask me every morning, I'd be like, oh, she's gonna ask me what I'm doing today. And I'd be like, I'm doing this, you know, I'm sending an email newsletter. And then from those two people we started a thing called Write Club. Do you know the book called Deep Work? About that time, it came out and we would sit around our desks congratulating each other on how many books you read about productivity and self-improvement and do no work. And then we would go and sit in the meeting room and say, should we just sit here for two hours and not talk? So, we developed this discipline together by sitting in the meeting room and not talking. And we'd just get so much work done because we would make ourselves sit in a library-like environment. And at the end of it, we'd be elevating, we like, oh my God, I've just written two emails. Sometimes we went to the Facebook page, and then that grew; now there are Write Clubs in five independent coworking spaces in London every weekday. As a result of that little group of people getting together back in 2014 and lots of people have come to that club over the years and published books and PhD papers and mums who their kids had left home and they'd always wanted to write, were like, I saw this on Meetup. I'll come into this coworking space, and I'll meet a group of people I may never have met before. And, you know, I'll write my book. The other one, this is one of my greatest hit stories, is

Urban MBA, which is a youth charity. And then in the pandemic, Fiona who fits out software and stuff in coworking spaces and one of her clients was, well still is, the Ministry of Sound; Urban MBA, ran that London Fashion Week show at the Ministry of Sound Coworking Space. And what people from the inside saw is that Fiona from the Ministry of Sound and Sophie from Urban MBA had a coffee, a breakfast event, and that a coworking space organized an event. And from the outside, people saw London Fashion Week, Ministry of Sound, global iconic music brands and Urban MBA, and there's one guy that he had been working with Urban MBA for seven years. He'd come to London Coworking Assembly things and just met people outside his circle. He came along and I saw the back of his website, and at the end of 2022, he passed a million in revenue as a result of walking into that thing. And then on that journey, different people have become involved in supporting his business, whether it was like another fashion show in a coworking space; different coworking spaces have supported that group of people on their journey. And maybe he would've made it anyway, but the intersection at that point of all these different people coming together in a coworking space and a group of people that they didn't have any commitment to each other. They were just in the same place and had energy and it made that happen. And there are endless examples like that.

A: One thing I really wanted to get across in terms of the transformative potential, I think the real key is zooming out and looking at it internationally as a connected larger picture. And again, what we're creating is this infrastructure of care. And so coworking, of course the action is all about the collaboration and the community and the cooperation. But the care is really what is holding that space; it's hosting and that labour, that emotional labour of hosting and gathering, but it is also interdependence and that builds resilience. And that is an incredible foundation like in a community, close to home, for thriving healthy people. And thriving healthy people are ridiculously powerful and impactful, and thriving healthy people create thriving healthy families, and thriving healthy businesses, and thriving healthy cities, and a thriving healthy economy, and communities, and neighbourhoods.

And so again, it's the ripple. It's the impact of coworking as a network, as an infrastructure of care, these networks of mutual aid. And again, it's that solidarity. It's the intention to be equals. But again, accessibility and inclusivity are key because if it is just privileged wealthy white IT workers gathering and traveling, we're gentrifying. And that has nothing to do with coworking. If it is a diverse range of people coming together in a local community to build resilience and to help each other be healthy and thriving and to feel like we're belonging, then we are building something impactful. And I think that will ripple for years to come. So, I think that's the potential that really excites me and where all the development is happening is around these sort of communities and networks of care. And I think that's a feminist principle. I think it's a feminist approach. I think a lot of these businesses are being run with feminist principles and approaches of equality and inclusivity and also those healthy thriving people are disruptive, and they organise, and they dismantle the status quo. And that's I think where we're really going to see the impact of transformative potential; it is to activate people.

B: Building on that a little bit before we go there's a question here. Why do we organize events and conferences? And one of the main reasons is so people can say things like that out loud with a group of people. So, in London we just did two events. One was about getting a more equal coworking space in an economic downturn and then we had one about inclusion on diversity for coworking space owners and local authority people and people that sell us software to all sit in a room and hear people saying things not dissimilar to what actually just said. You think coworking was a complete, you know, 1% coworking for the 1%, saying that that annoys people in London, and there's all this other like regular human beings from different ethnicities and underserved areas that are taking part in coworking. And it kind of dismantles that narrative that coworking is for posh white dudes with their start-ups. There is a freelancers' group, which is about 10,000 people on Facebook, and once a year Ed Goodman puts in "what do you think coworking is?" And loads of people from all over the UK launch in and say "coworking is where, you know, white dudes go to try and be the next Mark Zuckerberg" or something similar to that. They think that's what coworking is for. And then a load of people chimes in to say, this is what it is.

E: *Do you still see work and labour as the core of coworking or are we already beyond as you explained Ashley, it's more like the social infrastructure aspect.*

A: I think we're well beyond, and the people who are running social impact organizations and governments right now are recognizing the power and the essentialness of that infrastructure because they've been trying to create it with a top-down model for a long time, and it's been failing miserably. And really the reason for that is that we need to build it with those grassroots, that ground up collaborative approach. And I think that we continue to pursue models like cooperative ownership and employee ownership and all these different cooperatives and collaborative models in business and government. And so that we can continue that trajectory. I think there is a gap between coworking operators knowing that that's the business they're in and doing the work. They do the work intuitively, most of them, they do it out of the goodness of their heart and they do it in the way that I see leadership. For me, leadership is to serve and to protect in a very different way than we might attribute to say a police force, to literally serve. Leadership is to serve. And I think that many of those coworking operators, they do that intuitively. They serve their community, they're building resilience, they're breaking down things, the way things were built. But I think there's a big educational piece, and that's one of the reasons why I organize and bring people together is to make sure they understand how important it is that they are literally designing the future, not just the workspace. It's not just the new work landscape, it's the new landscape. And it might have taken a pandemic to bring people to the realization that we have the opportunity, we are standing on this, you know, tomorrow could be a brand-new difference, but we have the power, and we have the ability, and it falls on us to make that difference. That's true every minute of every day. And that's how I live. But a lot of people are coming to that realization now, and so we have a mass adoption, like a critical mass that we haven't had in the past. And I think the more we talk about the social impact and that other, all those other layers in addition to the work, changing the way of work is so essential to all of this. It's the easy in, but that's not at all what it's about. And so, I'd be happy to talk about coworking as work in the future.

It's what we do when we come together moving forward. So, I think the definition is definitely evolving. I would also like to list a couple other things. I think in terms of the promoters of the genuine coworking movement and spaces that I would suggest are operating that way would be the Centre for Social Innovation, the first coworking brand in Canada, 2004 Xpace. The Melting Pot, 312 Main in Vancouver. That's the project that we led that redeveloped the former police headquarters. And I think that's a fantastic example of how it has very little to do with working and very much to do with the community gathering, and the impact of the community there in the Downtown Eastside. I think that QWENCH in Victoria BC, the OpenHAUS in Portland, or even just reading their house rules, you'll understand that they really get the essence of coworking.

And then obviously I have to acknowledge Office Nomads in Seattle, New Work Cities in New York, you know, the first coworking spaces in these cities. Indy Hall in Philadelphia, The Foundery Buildings in Toronto, The Coop in Chicago. And these are all intentional, longtime coworking operators that are doing incredible work in their communities. And then I think in terms of policies moving forward and things we need to do at a governmental level to support coworking spaces. Toronto is a model upon which we could build. We have the model; we can point to it in other regions and say it works well here, and it can be adopted. But to have a policy to reduce property taxes for creative, cultural and non-profit coworking spaces or any membership-based collaborative space, to allow those employee-owned and cooperative models and to really take advantage of opportunities for community or cultural land trusts or coworking land trusts, including that ownership model, and then also for the government to take over some of the things that we are doing for independent workers, like healthcare. I don't want to run it, I've been running it for 10 years, for all the independent workers in Canada. The government should be running this, as universal healthcare but providing it for all members of the country. And I think that land trust and Indigenous participation and leadership are really at the core of all of that, going back to some of those principles that have existed for so many years, and we're paying for disrupting those. So, I think it's important to return to some of those principles around collaboration and stewardship and co-ownership.

E: *What motivates you to do all these things, to bring people together, to connect, to get engaged, to spend your time organizing independent workspaces?*

A: What motivates me to do this work is a drive to dismantle the status quo and to keep working towards social justice. I believe there is a better way, many better ways of doing things, that we can grow and create together, from the ground up. I believe that we all deserve better – a better work experience, and a better world. I want to see accurate representation and diverse leadership, I want to experience more inclusive teams, and more accessible workspaces, and I want to ensure my community maintains supportive networks of care and mutual aid so we can all maintain a more balanced and healthier lifestyle. I also believe I am a natural connector – a bridge builder and social translator – and I've always been able to meet people where they are at. My work in the world also brings people together because it's not just my work that I am doing. We are all working collaboratively towards a greater good. This is where the real power and strength of the coworking movement is. We believe in change, and we create it, together.

B: It's something I've always done. So, when I was at school, I would organise the school disco. When I was younger, when I was about 11, our house got burgled and me and my sister set up a neighbourhood watch. So we went up and down our road, we went to the police station, and you know "Oh, so you are two kids and you want to organise a neighbourhood watch, that's cute". And they kind of let us do it and we knocked on every door. And I think part of it is that my mum was a teacher, and she was always involved in scouts and guides and stuff like that and our church and the youth club and everything like that. So, it's kind of always been there. I find it very easy to get people together and other people are like "oh my God, you know, how does Bernie do that?". But I just don't have to think about hammering out event links to people and saying, "you've gotta show up to this". Janine Van Vanda Lindon, who runs the European CoWorker Assembly and is a partner in nine coworking spaces in the Netherlands, and I were a coworking in Europe once, and she told me, "I think your superpower is that you are the only person I know that can make people turn up to an event. And they are not sure why they are turning up to the event, but you get them to turn up, and then when they get there, something happens".

People sit down, and they are all a little bit awkward, and then they start saying "I need help with this". And a lot of those people, whether they run a big company or work on their own, they're looking for help, support, community camaraderie, something like that. And I have found over the years that when you sort of push them into a situation, they have to talk to each other, and then they start sharing. During the pandemic in London, every Tuesday and Thursday we did a Zoom call for any coworking space just to show up to, and sometimes 6 people would come and sometimes 50 people would come and they would just say "oh, this is really not working for me". Or they would announce "I have been trying to work out what to do with my space, but I have to close it". And then other people would say "we can help you close it and you can send your members here and we can buy that furniture off from you. And we know a lawyer who can help you shut down the business". So, at times of either triumph or devastation, other people are around to help carry that weight. And because of that peer-to-peer understanding going on, you don't have to retell the story every time. So, there are lots of instances like that that have come out and a lot of these things happen. I have been to a lot of conferences where four white guys stand on a stage telling you how amazing their business is. And they kind of accidentally exclude a lot of people. So we don't go around saying that all our events are mixed panels and diverse, we just do it. We just hope it will just become normal for people. A big topic for us at the moment is menopause in the workplace. And we just took that topic and put it front and in the centre of what we do. And loads of people are like "oh, thank God someone's talking about this". All these women are struggling in our coworking space, and we have no idea how to talk about it. And a lot of those topics I'm just not equipped to talk about. I just know the people to talk about them. So, we just organize things around that. Then, the urban MBA thing is a way of creating rather than like donating money to an online charity. The urban MBA thing is something that happens in your neighbourhood. We were already being involved in it, and when the whole George Floyd murder and Black Lives erupted, people were like "What can we do? We should tweet more". And we were like, actually, what you can do is you can come and make sure that people in your neighbourhood that are from the black community are supported by the work we do. And because it was so close, people could engage with it. So, not just as a result of George Floyd's murder, people became more alert, wanted to do something. And then the Urban MBA collaborated with a place called The Art Club in Hackney, which is a very ethnically diverse area. And they started running homework clubs and other people copied that. People were able to do something, to actually take action in their local community because of something they met through coworking, which is very different from sending money to Bob Geldof and hoping it will, you know, save a life.

E: *I think we saw something similar with how the coworking scene in Europe reacted on the Russian War on Ukraine. And with all the refugees coming in and opening up and providing help there.*

B: From the start of it, many colleagues with whom we have been connected through meeting in conferences over the last years thought that we should do something. And Ashley just said, "how are we going to do this?" And we developed a Coworking IDEA Project, broadcast if you like. And no one liked going into Zoom calls at that stage. And we had 70 people show up on three days' notice. Ashley curated it and Vika joined and broadcasted from her car and all these, I couldn't believe it. Also, in Warsaw there is a guy that worked for WeWork and is starting the Polish coworkers association and I've been to meet him in Warsaw, and he was like "I'm in Warsaw and you know, we are having these coworking spaces and hundreds of people are arriving in and we're putting them over here and people who run internet businesses are going to this co and people in Hungary and all these people I've met over the years from different places were doing something". What is also important is what we can learn from indigenous people. A few years ago, a software company accidentally published a comic book on the history of coworking, and it had indigenous people sitting around a campfire with laptops and people that are alert to indigenousness went mad. And I flicked through the comic, and I said "oh, I didn't really notice that", and my wife said "you know, as a white European male, you just wouldn't notice those things and it is, you know, actually quite offensive". And then we had an event on cultural appropriation. And I didn't even know what that meant back then; this was in 2019. I was like "what does cultural appropriation mean? What's that"? And then they explained the whole thing. And then a load of people who I've known for ages are like "I'm so glad someone's talking about cultural appropriation because we never hear about that. We just have to put up with this like, you know, dominant European economic protestant conversation and no one ever pays attention to what we say".

A: I think that the way we saw the coworking industry mobilize to support folks in Ukraine, and those fleeing Ukraine, was a perfect example of how essential this coworking infrastructure is. Our networks, internationally, and our organizations, our spaces, our members, our alliances – were all activated and collectively we were able to support, connect, educate, fundraise, protect and serve others who were in need. This happened quickly, and support was provided by peers, directly to those seeking assistance. These organizing efforts could now be replicated to support those managing natural disasters, war and other issues around the world. I think this is only the tip of the potential we have within these coworking alliances and networks.

E: *Where would you like to see coworking going in the near future?*

B: The short answer is the 15-minute city, but then when you say 15-minute city, that is already quite loaded. Do you know Town square? They have these two papers, one is called "What if Everyone was an Entrepreneur" and the other "What if everybody could walk to work?". Basically, they map out – they're not the only ones to do this and they certainly don't take credit for it – how a coworking space is. And this is happening already, and it's been happening for years. More coworking spaces that are part of the local community. So, people go there to flip open their laptop and get their work done and maybe there's an office there, but there's also a community area where, you know, people can do ballet. Like in England, more people used to go to the pub, more people used to go to church, and more people used to go to community centres. And community centres have been cut because of budgets. And particularly the conservative government is not interested in community centres. So, this coworking space is a multipurpose building where people from a local area come to work and gather, talk about local politics. In a city in the west of England, they run a model called Flat Pack Democracy. And what happened is about maybe 50, 10, 15 years ago, people who were active in local politics were sitting in a pub saying, "you know, by the way this council's going". And one of the counsellors who was from a political party said, to be part of a local council, you don't have to be part of a political party. And then, over a 10-year period, lots of independent people stood for council, and they won trust in the local community. And now basically everyone who runs the local government is an independent counsellor, and they manage that budget and it's turned into this thriving independent place that is not dependent on big party politics. It's dependent on what people in that area want.

A: Coworking in the future will just be called working. At least I hope so. I hope it becomes "normal" to work collaboratively, and intentionally. To build and maintain inclusive and accessible workspaces that support the entire human experience, not just our productivity. I hope the coworking movement manages to keep the values at the centre of the coworking industry and that we continue to inspire other industries and organizations to reimagine how and where they work. I hope that when we talk about coworking in the future, we all know we are talking about the action – the verb – not the space itself, but the intention. And I hope that instead of talking only about WeWork, people will start to acknowledge all the amazing independent operators and community organizers that are out there working hard, and changing the world, one neighbourhood at a time.

E: *Do you also see coworking as part of workers' associations, as cooperatives or something similar?*

B: Yeah, definitely. There's a place called Space4, which is in London, in Finsbury Park. They ran a series of workshops and one of those was about coworking in the local community. And all these people showed up, who had never been to a coworking conference in their life, but they were part of the kind of more local-type network. And they made us part of the class. We were like, oh yeah, we really value local communities! And then at one point they said, okay, so we, for our next exercise, we want you to go downstairs and talk to people in the street about coworking. And we were like, "I'm not doing that mate", you know, but that this is how you engage with people. I'm okay with talking about it, but I don't want to actually go and meet people. And we had to go downstairs and talk to people, and it was terrifying. And then we all came upstairs and had lovely stories to tell, and it was one of the best things we've ever done in our lives. But they got involved in challenging local problems, there was a big building project, and they were wondering what's this going to mean for the area. They go and look for all the small shops that English isn't their first language that are getting bullied by local developers and they go and support them, and they provide space for different types of groups. Every Wednesday they have this legendary lunch where you pay five pounds and lots of people from anywhere just sit around the table. Very often it's people from co-op tech activist groups sit around the table and have lunch together. They are part of a thing called One Tech, which is a UK-wide cooperative of cooperatives that are all kinds of tech and agency based. One of their legendary stories is they went up against a huge agency for this United Nations contract. And because they were a co-op, they got the job because of how their supply chain looked compared to the big agency. So, they were able to collectively win much bigger jobs than they would've done it on their own. And just that on its own is amazing. And it wouldn't be accurate to say that's a result of coworking, but I think the fact they understand coworking and its potential adds to them how they're going to achieve that.

Also, we ran this event called Coworking Nativo in Porto. Portugal is back at the ranch on tourism and places like Lisbon have become basically like Disneyland, and then more rural spaces are working out how to create another economy by attracting digital nomads. And then there's this tension of digital tourists coming in and like screwing everything up and then people coming who are actually committed to finding out more about culture and living and stuff like that. So that's not something I know about, but I'm just really experiencing more first-hand. But I think that's a very relevant conversation because we travel more and more, and people are able to work from a laptop and Covid just accelerated the future of work so much. So, now I can travel and work anywhere but what does that mean? And digital nomad visas are a very important part of how the future of work unfolds.

A: Coworking has always been connected to unions for me. My start in coworking came from the Ontario College of Art and Design where I worked as the Director of Finance for the Student Union. That's how I built my first coworking space – with and for the students – and so I've always taken the approach as a coworking leader that I do not "run the place", I am there to "serve the members". This community service also takes the form of fighting for freelancer rights; working collaboratively with the government to draft new policies; advocating for small businesses;

supporting social enterprises; encouraging entrepreneurship, co-operatives and employee ownership; and providing benefits and insurance options to allow people to work for themselves and still protect their families. In many cases our alliances and collectives act as unions, and this will only increase as more and more people become self-employed, contract or gig workers. Workers' rights will still need to be protected as we shift to remote and distributed work, and we are uniquely positioned to work collaboratively on tools and support for our members. We have many of these solutions in progress already.

E: *In Athens and in other places around Europe and beyond, we are experiencing this brand new co-living spaces popping around, along with a shift in coworking spaces' tenants profiles and the increasing numbers of digital nomads. Do you think that digital nomads, when they go to a place, when they visit coworking spaces, they got different interests, they've got that economic superiority. And, if they do it this way, does this undermine the possibilities of communities to emerge in coworking spaces because they have completely different profiles from the local coworkers?*

B: So, part of it has to do with cultural awareness and education. When I was 20 and EasyJet and RyanAir had just started, I remember going to Prague, and I was just amazed that I could buy a beer for 50p and the reason to go to Prague was not because it's an amazingly cultural place, full of history and you can visit these different buildings. It was like, I can get free beer, you know, the reason to go there was because it's cheaper and I can get drunk for less which is, I'd like to be more sophisticated, but that was it. And it's about intention, which is a bit hard to explain at scale, but there are communities of people who live together and do things together, and there's a lot of intention around it, whereas other places you see are like "come and live here and have cheap beer and get your work done". And it's how it's represented. I really think Lisbon sold its soul. It thinks it's going to boost the economy and properties are worth lots there, so if I bought a property in Lisbon in 2000, it might have cost me 10,000 pounds, and it's probably worth a million now. It's an advanced form of colonialism.

A: There will always be people trying to capitalize on the movement and make a profit. That is not what we are about. Coworking and coliving are about the "co". Digital nomads can travel and work respectfully, engaging with and learning from the locals. They can either contribute to or take away from the communities they visit. This is true for any traveller. We don't encourage people to take advantage of the resources in other communities as they travel. We want people to be mindful of gentrification, of the power and privilege imbalances in many cases. But we must remember that diversity is also a strength. Visitors are often welcome – this is how we share practices, network and learn from each other. The Coworking VISA program is a great example of that commitment – to accept coworking visitors and guests as our own members. But you also have a responsibility as a coworker. To be a good coworking member – you need to give to the community as much, or more than you take from it. It takes us all working together, in community.

Index

A
Affect, 4, 5, 19, 22, 38, 45, 46, 85, 91, 98–100, 107, 108
Affective work, 45
Affects, urban common, 18, 29, 30
Alternative organising, 51–53, 61, 63
Ambivalence, 10, 27, 46, 71, 113, 126–128
Assemblage, 1, 9, 84, 100, 155, 175–189, 197
Athens, 6, 8, 9, 69–74, 111, 116, 175–189, 239
Authentic sharing, 131, 134

B
Bankruptcy, 51, 52, 54, 61, 62

C
Care work, 8, 89, 90, 115, 130, 132
Class diversity, 20, 21
Collaborative, 8, 9, 34–36, 38, 45–47, 76, 113, 117, 122, 126, 129, 141–156, 175–189, 196, 197, 201, 204, 207, 221–225, 228–230, 233, 234
Collaborative labour, 38
Collaborative workspaces (CWSs), 6, 10, 67–78, 97–108, 125–128, 133, 134, 141, 162, 175–189, 207–210
Collective practice, 8
Commons, 5, 6, 10, 18, 24–27, 29, 30, 33–36, 38–42, 45–47, 61, 69, 74–77, 84, 86, 105, 114, 125, 127, 128, 132, 134, 135, 143–145, 150, 155, 162, 169, 180, 183, 191, 201, 209, 210, 212
Community, 1, 19, 33, 111, 126, 142, 161, 175, 191, 207, 221
Community economy, 194, 201
Community-led coworking (CLCS), 2, 4, 6, 8, 9, 125–134
Community management, 89, 112, 117, 118, 122, 167
Consumption, 24, 67, 68, 70, 72, 73, 75, 77, 78, 127, 148, 151

D
Digital nomads, 6, 67–78, 102, 160, 165, 166, 168, 222–224, 238, 239
Discourses on coworking, 19–22
Distribution of work, 129
Diverse economies, 9, 10, 101, 126, 130, 192–195, 197, 201–204

E
Embodied geographies, 8, 97–108
Emotion work, 115, 116
Employment, 5, 23, 24, 35, 42, 59, 62, 112, 113, 115, 117, 122, 143, 146, 147, 151, 162, 182
Ethics of care, 4, 7, 87, 90, 100, 101, 103–105, 108
Ethnography, 18, 51, 53
Exhaustion, 8, 126, 131, 132, 134

F
Feminist geography, 86
Finland, 6, 53, 163, 164

J. Merkel et al. (eds.), *Coworking Spaces*, https://doi.org/10.1007/978-3-031-42268-3

H

Host, 4, 7, 8, 28, 53, 54, 56–62, 69, 70, 73, 74, 86–92, 116, 117, 119–121, 129–132, 160, 163, 168, 169, 180, 181, 199, 209, 210

Hub, 10, 43, 118, 163, 165, 179, 207–216

I

Identity, 5, 25, 37, 69, 72, 75, 84, 85, 99, 103, 107, 115, 119, 126, 142, 155, 177, 178, 193, 194, 197, 207, 211–216

Idle capacity, 114

Inclusivity, 8, 97–99, 101, 103, 184, 232

Infrastructures of care, 7, 83–92, 100

Institutionalization, 142, 146, 154–156, 169, 177

Invisible work, 111–123

K

Koha economy, 101, 106–107

L

Lifestyle migrants, 68, 69

Local, 2, 3, 8–10, 33, 36, 38, 40, 43, 44, 46, 67–78, 83, 89, 91, 129, 131, 141, 143–156, 161–163, 165–170, 175, 176, 178–179, 182–185, 187, 188, 191–194, 197–201, 207–216, 224, 228–230, 232, 233, 236–239

Local development, 10, 90, 147, 162, 163, 166–170, 193

Low-paid service work, 122

N

Neighbourhood, 8, 70, 73, 131, 142, 144–156, 163, 175, 176, 180–188, 224, 227, 229, 230, 232, 235–237

Neo-municipalism, 142–146

New working spaces, 159–170

New Zealand, 8, 97–108

O

Organisational decline, 51–64

Organisational failure, 51–64

P

Participation, 2, 18, 29, 36, 38–41, 44, 45, 75, 76, 143, 145, 154, 163, 169, 170, 175, 179, 194, 195, 199, 200, 208, 234

Peripheral, 4, 10, 72, 145, 153, 155, 160, 163–165, 167–170, 193, 207–216

Policy, 3, 4, 9, 17, 39, 68, 69, 112, 141, 142, 144–148, 152, 154–156, 160–164, 166, 168–170, 178, 180, 187, 191, 192, 201, 203, 208, 227, 234, 238

Postcapitalism, 35

Postsocialism, 37–46

Post-Yugoslav space, 38

Potentials, 1, 17, 35, 120, 146, 163, 175, 192, 208, 221

Precarity, 10, 18, 19, 23, 24, 33–35, 42, 46, 47, 75, 77, 112, 113, 117, 127, 152, 210, 229

Presidi di Comunità, 9, 159–170

Public spaces, 18, 26, 27, 29, 37, 38, 40, 41, 43, 84, 162, 163, 178

R

Relational practices, 8, 111, 112, 114, 116, 117, 120–123

Remote working, 68, 170

Reproduction, 10, 42, 78, 112, 115, 122, 129, 133–135, 143, 153, 155

Resilience, 78, 170, 188, 204, 232, 233

Rethinking development, 216

Right to the city, 27, 29, 30

Rural areas, 2, 90, 160, 161, 165, 167–169, 191–193

S

Self organization, 5, 6, 34–36, 38, 40, 41, 44, 72, 125

Social and solidarity economy (SSE), 5, 9, 69, 75, 77, 141, 146–154

Social change, 2, 7, 10, 41–43, 75, 84, 87, 156, 207, 210, 215

Social economy, 148, 176, 180, 191–204

Social innovation, 1, 9, 73, 141–143, 155, 161, 162, 168, 176, 177, 194, 223, 234

South Working, 161, 164, 165, 168

Spaces of care, 86, 91, 98, 100, 101, 103–108

Spaces of possibility, 10, 129, 208, 210

Subjectivization, 154

T

Temporality, 74, 156
Tourism, 9, 68, 72–74, 118, 166, 167, 210, 211, 238
Transformation, 5, 6, 9–11, 21, 24, 26, 29, 34, 67, 72, 75, 77, 142, 145, 148, 176, 185, 186, 208, 216
Transformative potential, 1–11, 30, 75, 78, 98–101, 103, 108, 155, 186, 188, 189, 207, 208, 232
Turkey, 18, 20, 21, 27

U

University-community engagement, 198
University-community partnership, 192, 193, 198–200, 203, 204
Urban governance, 145
Urban movements, 141, 142

W

Work-life balance, 104, 164

GPSR Compliance

The European Union's (EU) General Product Safety Regulation (GPSR) is a set of rules that requires consumer products to be safe and our obligations to ensure this.

If you have any concerns about our products, you can contact us on ProductSafety@springernature.com

In case Publisher is established outside the EU, the EU authorized representative is:

Springer Nature Customer Service Center GmbH
Europaplatz 3
69115 Heidelberg, Germany

The manufacturer's authorised representative in the EU is Springer Nature Customer Service Centre GmbH, Europaplatz 3, 69115 Heidelberg, Germany. If you have any concerns regarding our products, please contact ProductSafety@springernature.com

Printed and bound by CPI Group (UK) Ltd, Croydon, CR0 4YY
28/09/2026
02224077-0004